COASTAL CONSERVATION AND MANAGEMENT

An Ecological Perspective

CONSERVATION BIOLOGY SERIES

Series Editor

Dr. F.B. Goldsmith
Ecology and Conservation Unit, Department of Biology, University College London, Gower Street, London WC1E 6BT, UK
Tel: +44(0)171-387-7050 x2671. Fax: +44(0)171-380-7096.
Email: ucbt196@ucl.ac.uk

Dr. E. Duffey OBE
Chez Gouillard, 87320 Bussière Poitevine, France

The aim of this Series is to provide major summaries of important topics in conservation. The books have the following features:

- original material
- readable and attractive format
- authoritative, comprehensive, thorough and well-referenced
- based on ecological science
- designed for specialists, students and naturalists

In the last twenty years conservation has been recognized as one of the most important of all human goals and activities. Since the United Nations Conference on Environment and Development in Rio in June 1992, biodiversity has been recognized as a major topic within nature conservation, and each participating country is to prepare its biodiversity strategy. Those scientists preparing these strategies recognize monitoring as an essential part of any such strategy. Kluwer Academic Publishers has been prominent in publishing key works on monitoring and biodiversity, and with this new Series aims to cover subjects such as conservation management, conservation issues, evaluation of wildlife and biodiversity.

The Series contains texts that are scientific and authoritative and present the reader with precise, reliable and succinct information. Each volume is scientifically based, fully referenced and attractively illustrated. They are readable and appealing to both advanced students and active members of conservation organizations.

Further books for the Series are currently being commissioned and those wishing to contribute, or who wish to know more about the Series, are invited to contact one of the Editors.

Books already published are listed overleaf...

Already Published

1. **Monitoring Butterflies for Ecology and Conservation**
   E. Pollard and T.J. Yates (eds) (Hb 1993 o/p, Pb 1995) xiv+274pp.
   ISBN 0-412-63460-0

2. **Insect Conservation Biology**
   M.J. Samways (Hb 1994 o/p, Pb 1994) xv+358pp. ISBN 0-412-63450-3

3. **Monitoring for Conservation and Ecology**
   F.B. Goldsmith (ed.) (Hb/Pb 1991, Pb reprinted four times) 275pp.
   ISBN 0-412-35600-7

4. **Evaluation and Assessment for Conservation: Ecological Guidelines for Determining Priorities for Nature Conservation**
   I.F. Spellerberg (Hb 1992 o/p, Pb 1994 reprinted three times) xvi+260pp.
   ISBN 0-412-44280-9

5. **Marine Protected Areas: Principles and Techniques for Management**
   S. Gubbay (ed.) (Hb 1995) xii+232pp. ISBN 0-412-59450-1

6. **Conservation of Faunal Diversity in Forested Landscapes**
   R.M. DeGraaf and R.I. Miller (eds) (Hb 1995) xxi+633pp, with colour plate section and colour foldout. ISBN 0-412-61890-7

7. **Ecology and Conservation of Amphibians**
   T.J. Beebee (Hb 1996) viii+214pp. ISBN 0-412-62410-9

8. **Conservation and the Use of Wildlife Resources**
   M. Bolton (Hb 1997) xviii + 278pp, with colour plate section.
   ISBN 0-412-71350-0

9. **Conservation Management of Freshwater Habitats**
   N.C.Morgan and P.S. Maitland (Hb 1997) ISBN 0-412-59412-0

10. **Tropical Rain Forests: A Wider Perspective**
    F.B. Goldsmith (ed.) (Hb/Pb 1998)
    ISBN Hb 0-412-81510-9, Pb 0-412-81520-6

11. **Grazing in Conservation Management**
    M.F. WallisDeVries, J.P. Bakker and S.E. van Wieren (Hb 1998)
    ISBN 0-412-47520-0

12. **Wildlife Conservation by Sustainable Use**
    H.H.T. Prins, J. Geu Grootenhuis and T.T. Dolan (Hb 2000)
    ISBN 0-412-79730-5

COASTAL CONSERVATION AND MANAGEMENT

An Ecological Perspective

by

J. Pat Doody
National Coastal Consultants
United Kingdom

SPRINGER SCIENCE+BUSINESS MEDIA, LLC

Library of Congress Cataloging-in-Publication Data

Doody, J. P.
Coastal conservation and management : an ecological perspective / by J. Pat Doody.
p. cm. – (Conservation biology series ; 13)
Includes bibliographical references (p.).
ISBN 978-1-4020-7248-2 ISBN 978-94-010-0971-3 (eBook)
DOI 10.1007/978-94-010-0971-3
1. Coastal ecology. 2. Nature conservation. 3. Ecosystem management. I. Title II. Series

QH541.5.C65 D66 2000
577.5'1 –dc21

00-046036

Printed on acid-free paper.

Dedicated to my wife, Jean

Extract from:
Knightly, 1982. *Folk Heroes of Britain*, Thames and Hudson.

The tide, however, rose in the usual manner and without reverence soaked the King's feet and legs. He, jumping back, then declared, "Thus may all the inhabitants of the earth see how vain and worthless is the power of kings." (Knightly 1982)

King Canute (c. 995-1035) King of Norway, Denmark and England

Contents

Preface

This book covers the major coastal habitats of nature conservation interest in the temperate regions of the northern hemisphere, and their management. It describes the present situation, with examples from the United Kingdom and Europe. It starts from the premise that the majority of habitats, and coastal systems generally, have been the subject of human use over many centuries. Despite despoilation through over-use and destruction, there remain important areas of coastline with considerable nature conservation interest. In one sense, left to their own devices, these areas are neither natural nor fragile, as is often depicted, but relatively robust habitats which survive in spite of human use. At the same time many of the areas considered to be of a high nature conservation value have been shaped, in part at least, by human activity. The purpose of this book is to promote and contribute to the conservation and conservation management of these areas.

It begins by describing the main coastal habitats and their management. Whilst the traditional approaches are considered, wider perspectives are introduced, which take account of the dynamic nature of many coastal areas. Recognising the significance of human use to the development and wellbeing of many systems, later chapters focus on modified habitats and the need for integrated management. In this context an attempt is made to link the sustainability of human use of the coast with the natural resilience of coastal formations to change.

The author's own experiences over 20 years are synthesised and linked with practical examples and scientific studies. The book also encompasses the views of a wide variety of individuals and organisations who have had a special concern for coastal conservation. However, the author is solely responsible for the content, interpretation and advice given here.

Acknowledgements

This book would not have been possible but for the generosity of the many friends and colleagues who have, over the past 20 years, shared their experiences and thoughts on the coasts and coastal wildlife of Europe and North America with me. I have spent many 'happy' hours (in rain, shine and heat, and at the mercy of midges and mosquitoes) tramping over dunes, walking along cliffs and occasionally viewing areas from the air and the sea. I should like to thank everyone; the nature reserve managers of the Scottish islands, officers from local, regional and national governmental and non-governmental organisations, stretching from Finland to Albania, academics and research workers everywhere especially in France, Greece and Turkey and to my many friends in the European Union for Coastal Conservation.

The Nature Conservancy Council (NCC), in particular Dr. Derek Ratcliffe, gave me the opportunity to specialise on coastal issues, for which I am very grateful. I should like to acknowledge his help and that of all my past colleagues, both there and in the Joint Nature Conservation Committee which I joined after the reorganisation of the NCC in 1991. I am indebted to the late Prof. Bill Carter for the insight he provided into the geomorphological workings of the coast. Without this I am sure that many of the concepts which are included within this book, and today form the basis for determining conservation action, would not have been so well thought through. Thanks also to Prof. Norb Psuty for his help and friendship and giving me the opportunity to work with him in New Jersey.

Stewart Angus (Scottish Natural Heritage), Richard Barnes (Cambridge University), Nick Davidson (Ramsar Bureau), Brian Ferry (Royal Holloway & Bedford New College), Prof. Alan Gray (Institute of Terrestrial Ecology,

Furzebrook), Duncan Huggett (Royal Society for the Protection of Birds), Mike Pienkowski (European Pastoral Forum) and Sue Rees (English Nature) provided comments on individual chapters. Lauri Nordberg (Finnish Ministry of Environment) contributed helpful comments and suggestions and Prof. Arthur Willis (Sheffield University) provided a much needed review of the final draft. Finally, the encouragement of Prof. John Packham was important to the realisation of this project. I should like to thank him for this especially, and also for his help in improving the text.

1. INTRODUCTION

1.1 Scope of the book

The coastal zone, defined in 1.4, stretches from the land to the sea and includes a series of habitats which occur either as individual units or in combination. This book concentrates on the conservation of those coastal habitats which cross the land - sea interface or form part an integral coastal unit. Following a general introduction to the principles of management and conservation, attention switches to individual coastal habitats (sea cliffs, saltmarshes, sand dunes and shingle structures), considering management issues specifically related to them and their associated biota. Discussion of the wider conservation issues associated with more complex systems such as estuaries and deltas are considered under a general heading of coastal wetlands.

Any book on habitat and ecosystem management is by definition really a book about human use and exploitation. This chapter sets out to provide the background to considering coastal management for nature conservation. It examines the historical impact of human activities on coastal systems and provides a broad definition of the coast and the origins of nature conservation. In Chapter 2 general definitions and principles of management affecting the nature of coastal habitats are considered. Traditional approaches to the management of key coastal habitats are linked to wider issues relevant to ecosystems. These embrace the concepts of integrated coastal management, considered later.

1.2 Early human occupation

From the moment that human societies moved from a hunter/gatherer role to one based on settlements associated with agricultural and pastoral use, the environment has been altered by their activities. Early settlements were often located near coastal areas, where there was a rich abundance of food, particularly in the waters of shallow estuaries and seas. The earliest occupants of this zone probably had little impact on the natural development of these areas accepting tidal movement, changes in sea level and storms as part of the price of living near to the sea.

As civilisations developed it would appear that some aspects of their activities had major, if unforeseen consequences for the way in which the coastal areas developed. Many of these effects would have been unrecognised as being in a sequence of cause and effect. For example, the loss of tree cover in the Mediterranean brought about by deforestation in the hinterland to provide timber for shipbuilding, had a major impact on the development of coastal sedimentary systems (Meiggs 1983). The early civilisations of the Minoans, and later the Romans, relied heavily on the sea; there is evidence to suggest that the growth of many deltas along the Italian coast can be attributed to this (Cencini et al. 1988). These changes are shown by the impact on a number of thriving ports such as the Roman harbour on the River Tiber (Figure 1.1) where sediment washed down by the rivers from the deforested hinterland helped restrict access to the sea.

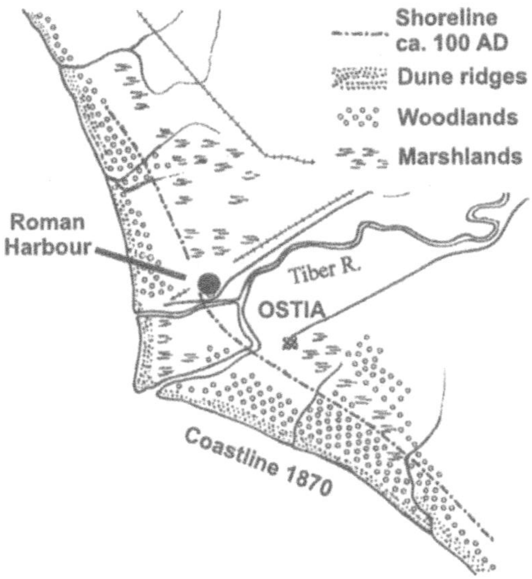

Figure 1.1. Growth of the delta of Rome, redrawn from Cencini et al. (1998)

Ephesus, one of 12 cities of Ionia (an ancient Greek district on the western coast of Asia Minor) was probably founded in the 11th century BC and became a major commercial centre in Roman times. Located near the modern town of Izmir, Turkey the harbour silted up between 750 BC and AD 726 (Eisma 1978) and was finally abandoned in the 14th century. Today the port lies some 3 miles (4.5km) from the sea. A similar process can be described for many other sites and continues today on the coast of Albania, where loss of forest cover in the interior has led to the growth of an extensive coastal plain in the northern half of the country (Figure 1.2).

This coastal plain has advanced seawards over the last two thousand years at a rate of 4m yr^{-1} due to the deposition of large quantities of silt from the surrounding hills (Shuisky 1983). The town of Lesha was an important port in the 11th Century, and now lies some 8km from the sea. This growth continues today with the destruction of mountain forests by fire and tree felling, and also poor maintenance of orchards and terraced lands which release sediment into the rivers (World Bank 1992). Land upheaval on the Finnish coast has had a similar effect. The medieval port of Ulvila (near Pori) is, for example, now several km from the sea.

Evidence, again from Italy (the Ombrone Delta on the northwest coast), shows a sequence of growth and decay which can be linked to different periods of human activity. The growth of the delta during Roman times was followed by erosion as depopulation occurred after the demise of the Roman Empire. A further period of expansion in the Central and Upper Middle Ages is linked to severe land exploitation when the delta grew again by some 2km. Another period of erosion coincides with population decline caused by the Black Death, which halved the population of Tuscany during the 14th and 15th Centuries (Innocenti & Pranzini 1993). With each decline there appears to have been an increase in woodland and a decrease in sediment movement from the land to the sea.

Figure 1.2. Sedimentary coastal plain derived from erosion in the hills of northern Albania; lagoons, saltmarshes and sand dunes

During these early periods of history the link between deforestation and the growth of sedimentary coastal systems may not have been recognised. However, as civilisations became more prosperous they undertook activities designed to control the natural forces of the sea. In Great Britain there is evidence of early embankment in the Severn Estuary in the south west where extensive Romano-British enclosure can be linked to wetland/intertidal occupation (Allen 1992).

1.2.1 'Primary' land claim

Loss of inter-tidal coastal land has increasingly been a feature of human use as civilisations become more prosperous. The Dutch have rightly earned the reputation for being great coastal engineers. Land claim in Holland, over more than 1,000 years, has extended its area by as much as 40%. In many of the major schemes both tidal flats and saltmarsh were included. In the Zuyder Zee, a total of 550,000ha were enclosed between 1920-1925 and throughout this century enclosure has continued in the Wadden Sea and southwest Holland (Dijkema 1984, Goeldner 1999). The techniques of enclosure, transferred to England in the early 14th Century, took advantage of the natural accretion of saltmarsh, and substantial areas of tidal land were 'won from the sea'. For example, new agricultural land was created in the Wash with 29,000ha, including 3,000 in the 20th century (Dalby 1957) and the Ribble Estuary, 2,000ha in the 19th century (Royal Commission on Coastal Erosion 1911) with further enclosures in the 20th century.

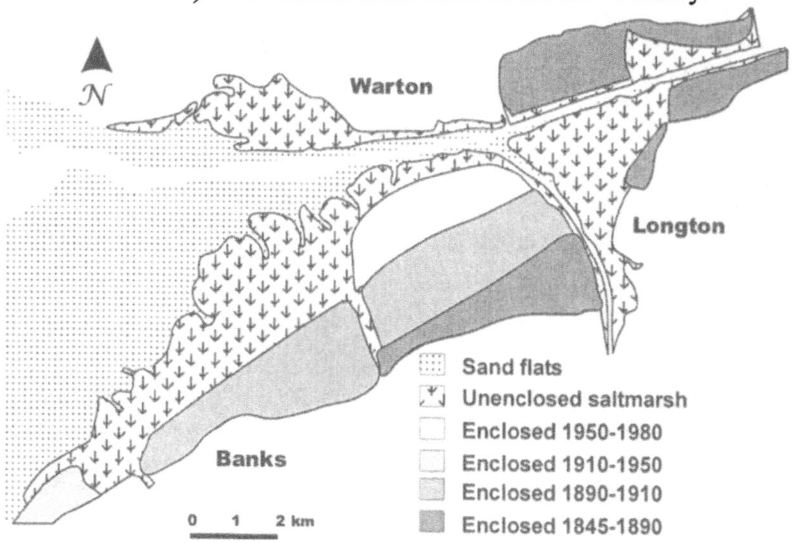

Figure 1.3. Agricultural land claim in the Ribble Estuary, Lancashire, England. The dates of reclamation are derived from Berry (1961) and the position of the 1855 coastline is based on the 1st Edition Ordnance Survey maps

Land claim on the Ribble Estuary shows the combined impact of direct loss of saltmarsh and the subsequent indirect loss of intertidal flats as new saltmarsh extends seawards. The development of the saltmarsh was aided by the planting of *Spartina anglica* in 1932. Figure 1.3 above, shows the pattern of enclosure and the most recent position of the edge of the saltmarsh.

A similar story can be told for France, where during the last century over half the saltmarsh of western France was enclosed (Géhu 1984). Géhu further reports that almost all saltmarshes in France suffer from deterioration and "nibbling", a process whereby small scale enclosure, including dumping of rubbish, slowly excludes the tide. In the Mediterranean, saltmarshes declined rapidly after the 1930s because of their extensive enclosure and conversion for cultivation (Goutner 1992). According to Sacchi (1979) most of the littoral lagoons which bordered the Italian coast have been lost, many artificially drained for agriculture.

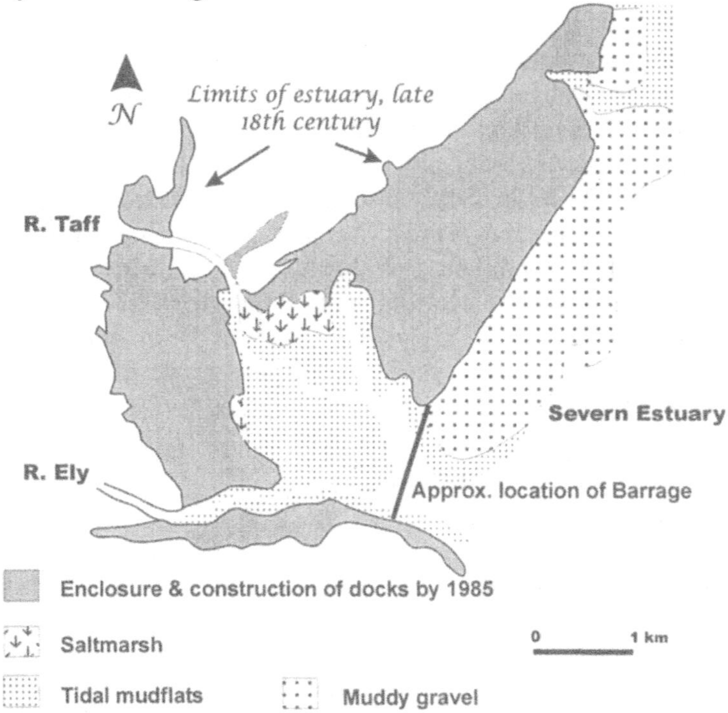

Figure 1.4. Loss of an estuary - industrial land-claim, Cardiff Bay, South Wales

Similar losses have affected other areas, especially those where industrial development is the main focus of human activity. On Teeside (northeast England), for example, almost all the estuary was enclosed with the loss of 84% of its saltmarsh and tidal flats to industrial and port development (Evans et al. 1979). In Wales the estuary of the Rivers Taff and Ely was finally destroyed when a barrage was built across the entrance to create a freshwater 'amenity' lake in Autumn 1999 (Figure 1.4 above).

1.2.2 Converted lands: 'secondary' land claim

Although loss of the habitat has been the predominant feature of human activity on the coast there are also examples where new habitats have been created. In their turn these have developed a nature conservation interest as, for example, in the early enclosure of saltmarsh in northwest Europe. Here coastal wet grassland, derived from the saltmarsh in Britain, includes some of the older grazing marshes behind sea walls. Many of these, together with their valuable plants (including species of high saltmarsh), rare invertebrate animals and birds, have survived until relatively recently. Rapid conversion of these coastal permanent pastures to arable land, which began in the 1960s, destroyed most of the nature conservation interest, as it had already done on long established pasture elsewhere (Doody 1995). This conversion completed the destruction of the tidal saltmarsh originally embanked and removed from the influence of the sea.

Enclosure of tidal land and lagoons for salt production is another widespread activity, especially in the Mediterranean. Traditional management of the salinas has also allowed development of their own special nature conservation interest; a range of endangered flora and fauna can survive such conditions (Walmsley 1995). More intensive salt production can reduce the wildlife interest but the use of tidal land for aquaculture can destroy it altogether. The result of this sequence, of progressively more intensive use, is ultimately to destroy any vestige of natural habitat and exclude most forms of wildlife.

1.3 Natural or man-made?

Many text-books describe coastal habitats in terms of their ecological or geomorphological development. Saltmarshes and sand dunes, in particular are described as exhibiting primary succession. Classic studies in Great Britain include those of Chapman (1938, 1941, 1959) and Steers (1960), covering saltmarshes and sand dunes on the North Norfolk coast and of Yapp et al. (1917) for saltmarshes of the Dovey Estuary in west Wales. All emphasised the natural status of the vegetation in areas where ecological processes appear to have been relatively free from human influence.

The vegetation was thus described as a series of community types progressing from the early pioneer stages to more complex forms related to the physical factors influencing their development. In the case of saltmarshes and sand dunes these include tidal, wave, wind, sediment and soil characteristics. This has led, in turn, to the impression that the process of succession takes place largely in a sequence which is determined exclusively by natural forces. Selection of areas for conservation designation is frequently based on the existence of recognised vegetation patterns based on the understanding of this 'ecological succession' and subsequent management often seeks to maintain this pattern in the face of change.

1.3.1 The importance of domestic stock

The view that coastal habitats are amongst the most natural must be tempered by the recognition that most, if not all, have been affected in some way by human action. For example, many apparently natural coastal systems have been influenced by agricultural use, notably grazing by domestic stock (Figure 1.5). Cliff-top grasslands and heathlands of north and west Europe, themselves often derived from the conversion of the original forest to more open landscapes, have been traditionally used as grazing pasture. In response to such use they have often developed a rich flora and fauna, which in the more exposed sites includes a distinctive maritime variant of more inland vegetation types. Examples include the maritime grasslands and heaths of the north, with species such as *Primula scotica* and the rich calcareous grasslands of the south and west where the warmth-loving *Ophrys sphegodes* grows with a wealth of other rare species.

Figure 1.5. Dunes in many temperate regions of the world have been grazed for centuries, Sandscale Haws, Cumbria, northwest England

Similarly the development of grassland and heath on dunes is often the result of their use for grazing. It is also clear that the open areas of upper saltmarsh previously provided rich pasturage in the winter months as they still do in many areas of Europe today (Dijkema 1984) and the early settlers in America cut hay on saltmarshes (Teal & Teal 1969). Throughout Europe most coastal areas have been used to graze domestic stock and as long as

8,000 years ago sheep herding was prevalent in many areas of the Mediterranean, especially in the winter.

1.3.2 'Natural' succession

Dunes and other coastal habitats may exhibit the characteristics of natural succession which appear to have developed without human interference. However, on closer inspection many such sequences appear to have been initiated by human activity. There are numerous examples of sand dune erosion due to overuse from such activities as cultivation and marram-cutting for bedding and thatch resulting in the major sand movement, see for example the Outer Hebrides, Scotland (Angus & Elliot 1992), Doñana National Park (Garcia Novo 1997) and Denmark (Skarregaard 1989). The growth of the 5km long dune spit of Bull Island, in Dublin Bay is directly attributable to the construction of a breakwater into the Bay between 1819-1823 (Jeffrey 1977). Given these examples, it is probable that few coastal habitats can be considered to be truly natural. The best that can be said is that the coastal areas especially the saltmarshes and sand dunes of Europe, exhibit the characteristics of natural systems (Doody 1999). This has important consequences for the way coastal habitats are viewed in relation to their conservation needs and is considered in more detail in each habitat chapter. See also Packham & Willis (1997, pp. 20-31) for description of the plant and animal community responses to environmental conditions.

1.4 Defining the coastal zone

Defining the coastal zone is fraught with difficulties. To some extent it depends on whether you look at it from the land or the sea. To the marine biologist the zone includes the sea and all those areas which are periodically covered by the tide. To the coastal ecologist the zone extends landward to the limits of tidal movement or the influence of salt spray on soils and vegetation. These limits may themselves be too narrow as the sea plays an important part in the development of essentially terrestrial habitats such as sand dunes. Here sand grains are driven onshore by a combination of tidal action, waves and wind. To the coastal geomorphologist the limits can extend as far as coastal agents of erosion and deposition operate and thus may include marine areas and whole river catchments. Biologists, concerned with the conservation of migrating birds using coastal habitats, may see their survival as being intimately bound up with their nesting habitat. In the case of some species such as wading birds, this may involve habitats as diverse as tidal mudflats, used for winter feeding, to Arctic tundra near the coast where breeding takes place. The geographical range may stretch from southern

Africa to the Arctic. The fishery biologist may view the coastal zone as a transition point between the Sargasso Sea and upland river beds where some migrating fish spawn.

For the purpose of this book something of all of the limits is involved. By their nature the zones immediately adjacent to the coast are indistinct, as tidal movement, storms and the effects of sea level change influence the relative position of each. Tidal areas (mudflats, sandflats, saltmarshes and transitions to brackish marsh, swamps and salt-influenced grasslands) are included, as are rocky shores, shingle beaches and cliffs subject to salt spray. Terrestrial habitats including sand dunes and shingle structures which are derived from the action of marine processes also form part of the sequence. Habitats occurring in combination within estuaries, deltas or lagoons (including areas claimed from the sea such as coastal grazing marsh) are also covered. Not included, except in relation to the geomorphological processes which affect the coast, are the sub-tidal waters and their plant and animal communities. The patterns and process which underlie the consideration of the conservation management of dunes, saltmarsh and shingle are the main focus of the companion volume (Packham & Willis 1997). This book takes a wider perspective of coastal habitats and ecosystems based on the following definitions of the limits of the zone from the land to the sea (Figure 1.6).

COASTAL WATERS		SHORELINE		MARITIME ZONE	
Offshore	Near-shore		Back-shore	Coastal habitat	'Terrestrial' habitat
Marine	⇐LW⇒		⇐HW⇒	Tidal Limits	
	⇐ Coastal processes ⇒			Limit of salt spray ⇒	

Figure 1.6. Definition of the 'coastal zone' (after Bird 1984). LW, Low Water Mark; HW High Water Mark

This definition does not take into account the influence of human actions which are of paramount importance in defining the zone for nature conservation purposes. This is dealt with in Chapter 2 and throughout the book, especially in Chapter 13.

1.5 Coastal landscapes

Coastal landscapes can be divided into two major divisions which provide a useful distinction when considering management options for nature conservation. The first is based on the relatively stable hard rock coast, where the geological structures are resistant to the erosive forces of wind, rain and the sea. Although more mobile sedimentary systems may be present in the matrix of cliffs and rocky shores, they are seldom extensive. The opportunity for new habitats to form is limited because of the lack of suitable mobile sedimentary material. This first landscape type can be further divided into two sub-categories depending on the nature of the surrounding land i.e. whether it has a high or low relief. The second type is dominated by soft rocks and abundant mobile sediments which are easily eroded and moved in response to tides, waves, wind and rain. This also includes two sub-categories, macro/meso-tidal coastal plains and micro/meso-tidal deltas and other coastal wetlands. The definition of the tidal range uses the familiar divisions quoted by Carter (1988) for spring-tides: namely micro-tidal <2 m, meso-tidal 2-4 m and macro-tidal >4 m. The above four very broad landscape types are themselves each composed of a series of habitats and are dealt with in more detail in the individual chapters. [Some of the habitats occur in several landscape types, and wider considerations reflecting their geomorphological characteristics are included.]

1.5.1 'Hard' rock coastal landscapes

A large proportion of the world has a rocky coast. These areas consist of low-relief coastal plains along "passive" continental margins to high relief, cliffed zones in "active" areas where continental plates have collided (Griggs & Trenhaile 1994). In Europe 'hard' rock coasts occur in both high and low relief areas where the underlying geological structures are resistant to erosion. They are usually characterised by low sediment availability. In areas of high relief the landscape includes steep cliffs, rocky shores, small embayments and pocket dunes with deep, usually clear offshore marine waters. They include the steep-sided fjordic landscapes in the north, and drowned river valleys (rias of southwest England, France, Spain and Portugal).

High relief areas exposed to the full force of the Atlantic form the typically cliffed, maritime coastlines (Figure 1.7). Salinity gradients occur in the soils on shore as salt spray driven inland results in vegetation transitions from assemblages with salt-tolerant species to typical inland types. The Baltic, Mediterranean and Black Sea coasts also include high relief and 'cliffed' coasts, but, because they are much less exposed, the impact of salt spray is more limited except in a few areas. As a result the plant and animal communities are less maritime than those in the north and west. Soft sedimentary structures such as pocket beaches, dunes and shingle shores occur but are usually small and restricted in their distribution.

Figure 1.7. Devonian Old Red Sandstone cliffs, north coast of Banff and Buchan, Grampian Region, northern Scotland

Low relief areas, particularly where isostatic uplift is occurring in the north, form more gently sloping landscapes. These include the small islands and shallow marine waters of southeast Norway and the skerrie coasts of southern Sweden. The shores may have special maritime features including narrow rocky shores and sandy/shingle beaches such as those that occur around much of the Gulf of Bothnia. The hinterland has more terrestrial habitats such as heath, scrub and woodland. The sloping karst landscapes of the Mediterranean, with their low-growing drought and grazing resistant scrub also fall within this category. A more complete definition of the cliff habitat includes the soft rock cliffs (glacial deposits, other Pleistocene sediments, chalk and clays) discussed next.

1.5.2 'Soft' coastal landscapes

'Soft' landscapes are complex and include a variety of both soft rock cliffs and sedimentary habitats. The essential difference between these and the harder rock types lies in the time-scale over which change takes place. 'Hard' rock landscapes may appear unchanging for decades or longer, whilst the softer rocks and sediments are more dynamic, altering in response to changes in climate, sea level, storms or tidal cycles.

Glacial deposits on land (including glacial cliffs) and at sea, together with some younger sedimentary rocks, such as chalk and the more recent Tertiary clays, can be eroded relatively easily to provide material for deposition elsewhere. Sediment is also derived from the hinterland and in non-glaciated areas may come from the erosion of mountainous or other areas of high relief. Over time, weathering produces large volumes of sediment which are then available for transportation and colonisation (Figure 1.8). This results in the development of extensive sedimentary plains with a complex structure including intertidal mud and sand flats, saltmarshes, other coastal wetlands and sand dunes. These often develop as low-lying land with gradual transitions from terrestrial habitats to intertidal and subtidal marine zones.

The nature of the sedimentary system depends on the balance between the energy from tidal movement, wave energy and river discharge. The two basic regimes (i.e. meso - macro-tidal and micro - meso-tidal tend to be 'tide dominated' or 'wave dominated', respectively) and help to determine the way these forces shape the coastal plains. They are described in a number of geomorphological texts (Pethick 1984, Bird 1984, Carter 1988, Carter & Woodroffe 1994). A brief summary of the way in which they affect the development of coastal areas is given below.

Figure 1.8. Eryngium maritimum (sea holly) colonising recent deposits of sand, Newborough Warren, Anglesey, Wales

Estuarine/barrier island coastal plains occur mostly in low relief areas with abundant sediment. Eroding glacial or other soft rock cliffs and river borne material, together with offshore marine sediments provide the sediment source. In macro to meso-tidal areas the sediment is driven landwards infilling river valleys, or accumulating behind barrier islands and other protective features. This results in the formation of the estuarine or barrier island coasts so typical of the southern North Sea (including the Wadden Sea). Habitats include the extensive intertidal sand/mud flats and saltmarshes, onshore sand dunes and shingle structures which may show successions associated with sediment deposition. Any hard rock cliffs are limited in length and lie within the softer sedimentary matrix. Salinity gradients within the estuarine waters are largely determined by interaction between tides and freshwater river flows.

Deltaic coastal plains occur adjacent to areas of both high and low relief and with 'soft' and 'hard' rocks - in micro-tidal areas. Tidal influence is limited; storms and longshore drift provide the main driving forces for accretion and erosion. Much of the available sediment is derived from erosion in the hinterland, often following deforestation. River-borne sediments are washed offshore to form typical deltaic plains which project beyond the coastal margin. These systems include saltmarshes, sand dunes and lagoons. They are typically found in the Mediterranean and Black Seas and in the southeast Baltic where the tidal range rarely rises above 1m on a spring tide. Salinity gradients are less pronounced than in macro-tidal areas and hence the vegetation and associated animals tend to occur as mosaics without obvious successions and zonations.

Although this represents a very simplified view of an often complex mixture of habitats occurring in each coastal landscape type, the four broad categories do provide a first order classification and a useful distinction for management purposes. The harder resistant coastline poses a very different series of management questions to the soft sedimentary structures which are dynamic and prone to more rapid change. In the first, change is slow and losses of habitat are usually permanent. In the second, change may be rapid and new habitat can be created if the conditions are suitable for the deposition of mobile material. Thus the conservation needs of ecosystems, individual habitats and sites within each landscape type require different approaches.

2. GENERAL PRINCIPLES

2.1 Origins of coastal conservation

The massive changes in land-use and habitat loss described in Chapter 1 have resulted in the remaining areas of unspoilt coastal landscapes being considered to be particularly precious. As a result nature conservation organisations world-wide have expended considerable time and money in an attempt to prevent the worst excesses of human destruction. Despite this, loss of habitat and degradation of the environment continues. This chapter reviews some of the general principles of habitat, species and site conservation management as they apply to coast.

2.1.1 Species and habitat protection

During the formative years of the conservation movement protection of individual species of birds was an important consideration. The early Acts of Parliament in Britain were mainly concerned with bird protection: The Sea Birds Protection Act, 1869; The Wild Birds Protection Act, 1872 and The Wildfowl Protection Act, 1876. Species protection remains an important component of the legislation and prominent among more recent legislation are the Protection of Birds Act 1954 (amended in 1967) and the Conservation of Wild Creatures and Wild Plants Act, 1975.

In the early post-war period in Great Britain, considerable interest was expressed by the leading ecologists of the day in the wider concepts of wildlife conservation. This is perhaps best described in A.G. Tansley's book "Our Heritage of Wild Nature" published in 1945. The thesis of this book is that "the preservation of rural beauty" must involve the conservation of much of our native vegetation. At that time it was recognised that the planning of this work must "be balanced and harmonised with land utilisation and for agriculture and forestry". These wider needs were recognised in the National Parks and Access to the Countryside Act, 1949, but only for a few National Parks were established. The nature conservation elements of the Act were mainly concerned with the establishment and management of nature reserves and the designation of Sites of Special Scientific Interest (SSSIs) as was the Wildlife and Countryside Act, 1981, which strengthened site protection measures.

In America threats to populations of herons and egrets led in 1903 to President Theodore Roosevelt signing an order to protect these species and others on Pelican Island in Florida. This effectively provided the forerunner to the establishment of the first wildlife refuges which today include the 440 sites making up the National Wildlife Refuge System. Other protected areas include a wide variety of National and State Protected sites and voluntary reserves. The Wellfleet Bay Wildlife Sanctuary is one such, owned and managed by the Massachusetts Audubon Society, part of the United States 'Audubon Alliance' and lies adjacent to the Cape Cod National Seashore (Figure 2.9).

Figure 2.9. Wildlife reserves form the backbone of nature conservation effort in the US, as elsewhere in the world. Wellfleet Bay Wildlife Sanctuary, Cape Cod, Massachusetts

In the rest of world the pattern of development of nature conservation practice has been similar. National parks and nature reserves provide the cornerstone for the conventional approach to terrestrial nature conservation by the protection of the most important sites representative of the remaining natural and semi-natural habitats and important species concentrations (see, for example, International Union for the Conservation of Nature 1992 for an overview of nationally protected areas throughout the world).

In recent years the increasing recognition of the need to conserve a series of sites for the protection of migratory animals has led to a number of international conventions and agreements. The Western Hemisphere

Shorebird Network, a collaborative venture between private and governmental organisations and launched in 1985, covers north and south America. It stretches from the Arctic to Tierra del Fuego and is an indication of what can be achieved between nation states.

This world-wide approach has been taken even further within the United Nations Environment Programme, Rio Convention, agreed in 1992 which puts conservation management firmly in the economic and cultural arena. From this perspective nature conservation can be defined by a philosophy which embraces the principle of wise-use of the environment which does not compromise the ability for future generations to enjoy its richness and diversity. By this definition the preservation of biodiversity and naturalness are part of a spectrum of conservation aims embracing human activities which are sustainable and firmly linked with environmental principles.

2.1.2 Assessing nature conservation importance

Traditionally the nature conservation value of individual sites has been measured against a suite of attributes which help define the 'naturalness' and hence conservation importance of the areas concerned. Ratcliffe (1977), for example, defines the selection criteria for nationally important sites in Great Britain under 8 headings:

1. Intrinsic appeal;
2. Diversity of habitat and species in one place at one time; the traditional view upon which site selection is often judged;
3. The size and rarity of species concentrations;
4. Structural diversity;
5. Temporal dynamics between habitats, habitat mosaics and species movement. Included within this may be the movement of species within and between habitats, successional characteristics;
6. Naturalness;
7. Recorded history;
8. Research.

At a European level the European Union Habitats Directive (Directive 92/43/EEC on the conservation of natural habitats and of wild fauna and flora) provides for the establishment of a network of protected sites across Europe which will be known as 'Natura 2000'. This network will include sites designated as Special Areas of Conservation (SACs) under the Habitats Directive and Special Protection Areas (SPAs) under the Birds Directive, which is already in place. The concept of 'Natura 2000' has at its centre the idea of conserving habitats and species of importance at a European level in a "favourable conservation status". The initial selection for Candidate

Special Areas for Conservation - 'Stage I' of the process adopted within the Directive - is done at national level. The criteria for site selection at a European level 'Stage II' clearly give expression to the importance of developing networks of sites at a wider scale based on biogeographical regions. These 'Stage II' criteria are listed in Annex III of the Directive and can be summarised as:

1. Relative value of site at a national level;
2. Relationship of the site to migration routes or its role as part of an ecosystem on both sides of one or more Community frontiers;
3. Total area of the site;
4. Number of Annex I habitat types and Annex II species present;
5. Global ecological value (overall assessment, based on 1-4 above) of the site at the level of the Biogeographical region and/or the European Union as a whole.

Setting individual habitats within a wider context helps determine rarity values as well as links between sites important to mobile species. Spatial variation of habitats and species thus forms an important part of the assessment of the nature of the coastal resource. This means determining how frequently a particular habitat occurs, over what geographical range and how large are the representative examples. This variation can be considered at a number of scales (Figure 2.10).

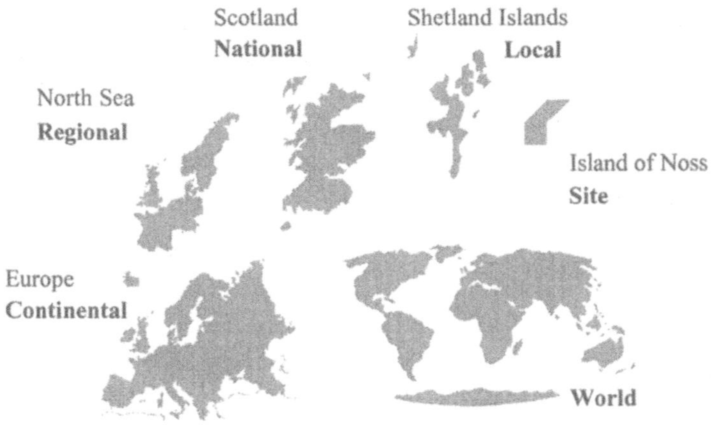

Figure 2.10 Geographical scales for habitat and species conservation

2.2 Coastal Management

Management implies some form of intervention towards a desired goal. The Oxford English Dictionary includes "wielding" (tools), "control" (household or State), "take charge of" (cattle etc.) or "subject to one's control" (animal), all of which deal with human manipulation of the environment. Although natural forces often define the character of the coast, its conservation significance and the uses to which it can be subjected, most coastal ecosystems have been influenced in some way by anthropogenic activity. Indeed the maintenance of natural environmental quality, which helps to define nature conservation, in many instances requires human intervention.

2.2.1 The traditional approach to management

Management for nature conservation on the coast today must be set within an historical perspective, where the protection of a rare and diminishing 'natural' asset is paramount. Having identified the area of coastal conservation importance and secured its protected status (possibly as a nature reserve), the next stage is to establish a management plan aimed at conserving the features for which the site has been selected. The first stop for the conservation manager will be to identify the ecological basis of the coastal habitats and the species for which the site is established. Ecological textbooks will frequently include a diagram showing the first stage in the development of a sand dune, beginning with the accumulation of sand over an obstruction on the beach (Figure 2.11). From this point sand accumulates until eventually it becomes stable enough to support heath, scrub and woodland in the absence of human interference.

Figure 2.11 Ammophila is specially adapted to help 'initiate' succession on a beach

Saltmarshes are also often depicted as progressing along a reasonably predictable path. Here ever more complex plant communities are postulated as sediment accretion increases the height of the marsh; see for example Chapman (1964) and Long & Mason (1983). Eventually the marsh reaches a vegetation climax where salinity effects are minimal. This may include scrub or even woodland, but succession often ends with brackish swamp vegetation.

Such successional processes are usually inferred from the spatial relationships of the vegetation at a given point in time. The coastal conservation manager may readily accept change as part of the normal course of habitat development, particularly in the early stages of succession. However, when this change involves the growth of invasive taxa such as *Spartina* spp. on saltmarshes or *Hippophae rhamnoides* on sand dunes, at the apparent expense more natural and perhaps rarer plants and animals it is often considered to be deleterious to nature conservation interests. Preventing such change, which may include arresting succession in older more stable habitats, is a process familiar to most coastal managers. This will be considered in more detail in the individual habitat chapters, but at this stage it is important to recognise that this relatively simple picture of succession can result in the employment of scrub control, a major and expensive management tool, when other more effective options are overlooked.

At the other end of the management spectrum, change, as manifested by erosion is often seen as a threat to a variety of habitats. Mobile sand in mature dunes will elicit a desire to 'protect' them and considerable energy may be expended on building sand fences and controlling human impact. Separating what is really damaging to a habitat from more natural forms of change, essential to their development, can be difficult. This may be especially true when the conventional wisdom from other quarters, such as engineers, reinforces the view that the coast requires protection. However to be truly effective management must include a wider perspective of what might be appropriate.

2.2.2 Coastal complexity

A more critical look at the natural world, shows the generalised picture of succession is not a true reflection of the real world. Saltmarshes are much more complex systems than the straight forward succession sometimes described in early ecological texts suggested. Accretion does result in more diverse forms of vegetation and associated animal species and transitions to terrestrial vegetation can be the most complex. But a series of steps can form as new saltmarsh develops to seaward of the eroding cliffed saltmarsh. Salt pans represent another element in the complex mosaic and deposits of seaweed on the tide-line may smother the surface vegetation creating further spatial variation as the strandline deposits rot. Sea level change may have long term implications for the location of the upper saltmarsh communities as the zonation is pushed landward or seaward depending on whether relative sea level is rising or falling.

Dunes show primary succession with the accretion of sand which is aided by specialist plants (notably *Elytrigia juncea* and *Ammophila arenaria* in northwest Europe). However, once the main body of the dune is formed other processes come into play and the development from mobile foredunes and yellow dune to grassland, heath, scrub and woodland is not a straightforward progression. Blowouts occur with or without the intervention of man and can be the precursors of dune slacks (Ranwell 1972a). Similarly the reprofiling of dune ridges under the influence of changing wind patterns brings an infinitely variable topography, the origins of which may be difficult or impossible to unravel.

The nature conservation interest of sea cliffs composed of soft rocks is also dependent, in many instances, upon topographic change. A degree of instability is essential to the maintenance of open, species-rich communities at a number of sites. Slippage, especially when associated with the freshwater run-off, may also be important for a wide variety of invertebrates. Figure 2.12 gives a picture of some of the more significant interactions associated with sediment and water movement, key elements in the development of coastal systems.

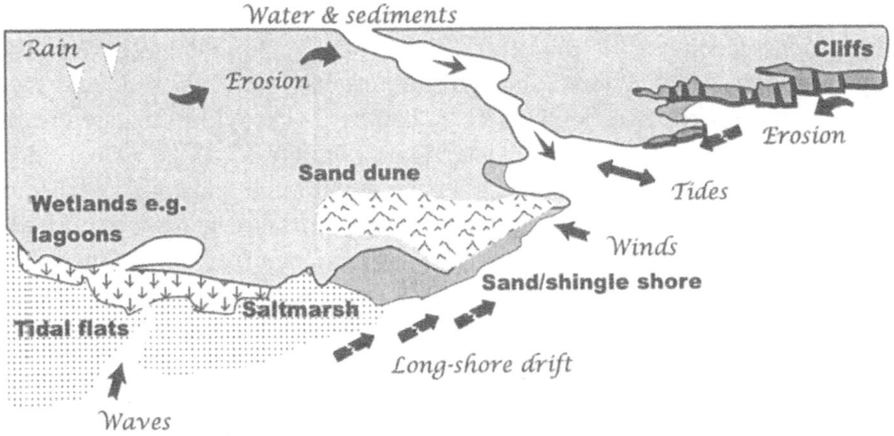

Figure 2.12. Some key elements of a 'natural' and 'dynamic' coastline

2.2.3 Human influences

Throughout history humankind has influenced coastal systems. In addition to the whole-scale changes accompanying land-claim through drainage, agricultural use and infrastructure development, other influences are significant. Each of these interferes with the functioning of the system, reducing its ability to react to the natural forces acting upon it. Some of the key ways in which the coast is affected are shown below (Figure 2.13).

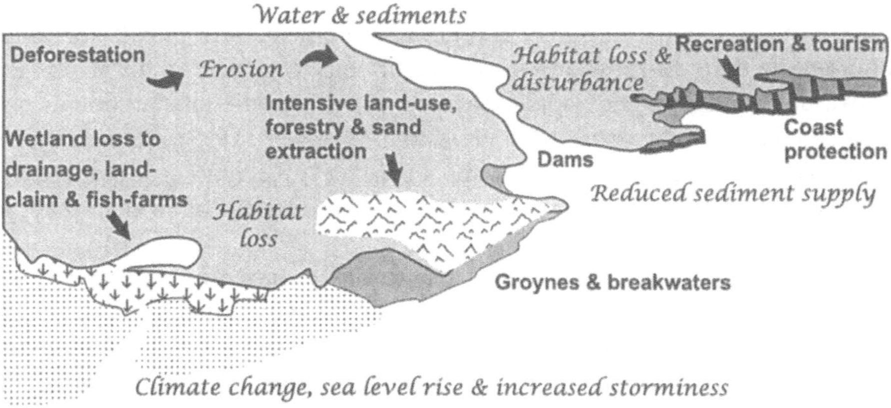

Figure 2.13. Some key aspects of habitat loss and interference with coastal processes

2.2.4 Coastal change and conservation management

Coastal processes (ecological and geomorphological) operate at a number of spatial and temporal scales. Individual habitats show successions and transitions which react to the natural forces acting upon them. These are built upon species interactions and adaptations as well as soil and water conditions; Packham & Willis (1997) chapters 2-3 provide a detailed review. Human activity can alter these systems in a variety of ways. At one level a change in grazing regime may affect the structure and species composition of saltmarsh vegetation. At another, the shape of an estuary is modified by enclosure, which in turn may change tidal flows and sediment patterns. The conservation of mobile species, such as migrating birds or fish, are also affected as different management objectives can make the areas more or less suitable for individual species.

The movement of barrier islands, sand dunes, tidal marshes and swamps is a natural response to the forces of the sea. The balance between erosion and accretion of the flat low-lying landscapes, around estuaries, has proved a flexible response to changing sea-levels since the last Ice-age. However, historically the coastline is also the zone where the first attempts to 'control' nature were made. Nature conservation management, in developing strategies for conserving species and sites, has often taken action designed to protect the surviving interest in the face of change. As will be discussed below this may not always be in the best long-term interests of either the habitats or the species being 'protected'.

Later chapters are mainly concerned with a description of the nature conservation interest of a variety of habitats and ecosystems. They also provide a practical guide to management in coastal locations. In keeping with the way in which conservation management of sites, species and ecosystems has developed, they deal first with the principal coastal habitats, combinations of habitats and species populations. Each of the habitat chapters describes the main conservation features and looks at management and management practice in relation to the competing demands for natural resources from human activities. Secondly the book looks at new approaches involving more integrated forms of management - between habitats, sites and people - and how these might provide a more certain future for the coast and the maintenance of biodiversity. In addition, later chapters consider the way management decisions on individual sites may influence the conservation of ecological networks (of migrating species) and the implications for wider initiatives. Coastal zone management and the importance of social and economic integration are also touched upon. A summary of the main coastal components and the major landscape types as they relate to the chapters in this book, is given below (Table 2.1).

Table 2.1. Coastal landscapes and the chapters where the habitats/species interests are covered

Primary Landscape Type	**Components**	**Chapters**
Hard rock high-relief cliffed landscapes	Sea cliffs, sea stacks & islands	Chapter 3 Sea cliff vegetation Chapter 4 Seabirds, sea cliffs & islands
Hard rock low-relief landscapes	Rocky shores; skerries & rias	Chapter 10 Coastal wetlands
Soft rock cliffed landscapes	Unconsolidated earth and clay cliffs	Chapter 3 Sea cliff vegetation
Soft coasts macro/meso-tidal coastal plain	Estuaries Mud flats; sand & shingle beaches, & shingle structures	Chapter 10 Coastal wetlands Chapter 5 Saltmarshes Chapter 7 Sand dunes Chapter 8 Shingle beaches & structures Chapter 9 Dungeness Chapter 11 Grazing marsh
Soft rock coasts meso/micro-tidal coastal plain	Deltas & lagoons	Chapter 10 Coastal wetlands Chapter 5 Saltmarshes Chapter 7 Sand dunes Chapter 11 Salinas & rice fields

Throughout an attempt is made to provide an historical perspective as well as consideration of the wider social, political and economic implications of different forms of management. Though not the primary focus of this book, coastal development, in the face of global warming is also discussed. This is especially important when so much of the human population is concentrated on or near the coast. In many areas this has caused the almost complete obliteration of the coastline leaving little or no margin for accommodating storms and/or rising sea level (Figure 2.14). This has important consequences not only for nature conservation, but also the sustainability of human use whether for commercial or leisure purposes.

Figure 2.14. Tourist urbanisation often leaves little or no natural coastal margin, Costa del Sol, Spain

3. SEA CLIFFS & SEA CLIFF VEGETATION

3.1 Introduction and scope

Sea cliffs create some of the most attractive and spectacular coastal landscapes in the World. Their distinctive nature is defined by the extent to which the material forming it is consolidated ('hard' rock cliffs), developing in resistant bedrock and unconsolidated ('soft' rock cliffs), developing in easily-eroded materials. The former are stable though they may be steep and inaccessible, the latter unstable and highly erodable. These attributes help maintain freedom from intensive human use, which is an important element in the survival of sometimes considerable nature conservation interest. The rock types 'hard' and 'soft' lie at the two ends of a spectrum which support very different nature conservation interests.

'Hard' rock cliffs made of older resistant rocks such as those depicted in Figure 1.7 (above) tend to have an apparently stable, if thin, soil on ledges and slopes. The nature of the vegetation is determined by the degree of exposure to salt spray and the acidity of the underlying rock type. Cliffs exposed to oceanic swell, onshore winds and storms have a maritime vegetation dominated by salt tolerant plants, some more typically found in saltmarshes. As the influence of sea spray diminishes progressively more inland types of vegetation occur. Exposed slopes out of reach of the direct influence of sea water and salt spray, support acid grassland and heath on acid rocks and calcareous grasslands on chalk and other limestones. On less exposed cliffs a variety of scrub and woodland communities exist. The steepest slopes may be devoid of vegetation, or nearly so, and provide ledges for nesting sea birds safe from predation (Chapter 4).

'Soft' rock cliffs are much more unstable. In the most easily erodable rocks (such as Pleistocene glacial clays and sands) slippage may be frequent. This is especially true for areas where marine erosion removes the 'protective' beach built up through weathering and slope transport processes. In such situations the slopes have little or no vegetation and animal life is limited to species requiring open habitats (including notable invertebrate fauna). Some cliffs may be protected by a beach and/or have a greater physical coherence. Here the slopes exhibit less frequent movement and despite periodic, and sometimes massive, slumping can nevertheless become stabilised with grassland, scrub or even mature woodland. Nesting sea birds are unusual as there are seldom any suitable ledges. Between these extremes an infinitely varied series of habitats occur which can be amongst the most natural and least influenced by human use of any coastal area.

This chapter reviews the important characteristics of sea cliffs including the management of cliffs and cliff-top vegetation. An attempt is made to distinguish between issues affecting the nature conservation value of 'hard' and 'soft' rock cliffs. There are few detailed ecological accounts of this habitat, but Larson et al. (1999) consider all cliffs, not just coastal, and Rodwell (2000) describes sea cliff vegetation.

3.1.1 Habitat definition

Sea cliffs are defined by the nature of the slope and contact with the sea. Hence sea cliffs develop wherever erosion of the base of the cliff occurs or where removal of accumulations of debris from above takes place through wave action. A simple definition of a coastal cliff is one which is:

'formed at the junction between the land and the sea where a marked break in slope is formed by slippage and/or erosion by the sea.'

3.1.2 Habitat type

The way in which cliffs are formed depends on the nature and resistance of the underlying rock, and the degree of exposure to processes of weathering including rain and wind, frost and attack from the sea. The combination of marine and sub-aerial processes promotes instability in the rock face, the most important factor in the development of sea cliffs and their vegetation. Even the hardest rocks can be eroded, though slowly, as rain and frost cause erosion along joints and faults, to create crevices, notches or ledges. Softer rocks, such as boulder clay cliffs, show mass movement of entire slopes as marine erosion removes the protective beach. Erosion is intermittent and unpredictable, depending on the exposure to wave attack and the structural strength of the rock (Griggs & Trenhaile 1994).

The older harder rocks of the north and west of Europe are the most resistant and help to form the spectacular vertical or near vertical cliffs, often of granite, which abound there. Erosion here may be imperceptible. By contrast the much younger and softer rocks of the chalk cliffs may erode much rapidly. In the case of the Severn Sisters in Sussex this may be at a mean rate of up to c.0.5m per year (May 1977). Locally even more rapid rates of erosion occur on the boulder clay coastline of Holderness where average rates of 1.8m per year have been reported (Lee 1995). Here whole villages have disappeared in historic times and buildings continue to fall into the sea today. Rates of approximately 1m per year have been recorded on the Pleistocene cliffs of Poland (Subotowicz 1994). On the Black Sea coast of the Ukraine eroding cliffs of sedimentary rocks in loess, clays, loams, shelly limestone, sandstone and shale have rates which range from 0.5 up to 6.0m per year (Shuisky 1995).

Carter (1988) identifies four major cliff types which are shown below in sections A-D (Figures 3.15 and 3.16). These are listed in order of their degree of 'hardness' to emphasise the importance of stability rather than form to the nature conservation interest. A, B and C are essentially the 'hard' rocks referred to above, D the 'soft' rock category.

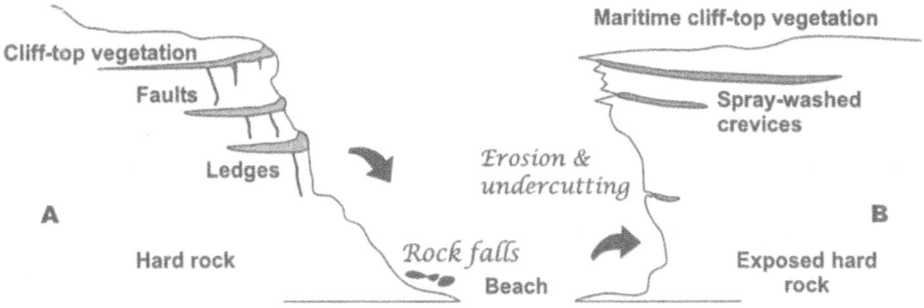

Figure 3.15 Hard rock cliffs - A - jointing & fault erosion and B - undercutting

A. Cliffs in highly resistant rocks e.g. harder limestones, basalts and granite where erosion is slow and movement results from splits along joints and faults. They are usually vertical in form and may include blow-holes, platforms and stacks.
B. Resistant but exposed rocks where weaknesses result in notches and overhangs. Erosive forces include wave attack, sea-water spray, wind and rain.

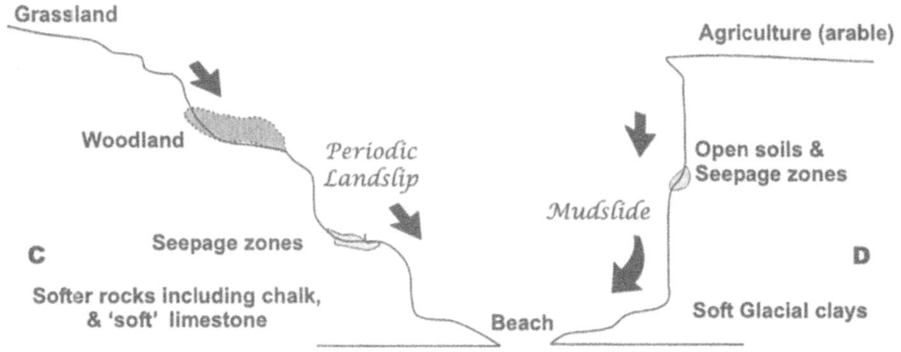

Figure 3.16. Resistant & soft rock cliffs - C - rotational slips; D - soft rocks, rapidly eroding in response to weathering and marine erosion

C. Softer rocks where undercutting creates pressure resulting in rotational slips. These may be large or small and can occur in limestones and shales which are more prone to erosion than A and B above.

D. Rapidly eroding 'soft' rocks - mudstone and glacial material. Glacial deposits and some younger sedimentary rocks such as chalk and more recent Tertiary clays can be eroded relatively easily.

3.2 Habitat distribution

Although sea cliffs are distributed around the coastlines of the world, Pethick (1984) tentatively suggests that they occur more frequently in higher latitudes. Here erosion is brought about by failure of the overlying rocks as a result of the action of rain and frost and undercutting of rocks and talus slopes through high wave energy, which act in combination. This is a very generalised picture but it does emphasise the role played by wave energy in cliff formation.

3.2.1 Habitat distribution in Europe

Maritime sea cliffs are distributed throughout Europe and the exposed Atlantic coasts of France, Spain and Portugal all have important examples. Sea cliffs also occur in the enclosed seas of the Baltic, Mediterranean and Black Seas. A summary of known distribution is shown in Figure 3.17 derived from several sources including van der Maarel (1993).

The cliffs range from the highly indented 'hard' rock coastlines of northern Norway, the western coasts of the British Isles, northern France (in particular Brittany) and northern Spain and western Portugal. The rock types range from the hardest granites (most prevalent in the north), rising to 426m on the island of St Kilda off the coast of northwest Scotland to the limestone cliffs of the Channel coast (up to 100m) and the megacliffs of the northeast Adriatic which may reach 1,200m where the Velebit Mountains meet the sea. Limestones and shales occur in the west (including western Ireland, Wales and southern England and France (around Cape Blanc-Nez). The softer Tertiary chalk and clays of southern England and the Pliocene - Pleistocene glacial deposits of the east coast of southeast England, west coast of Denmark and the southern shores of the Baltic (30m) and the sedimentary rocks of the Algarve (up to 30m) present some of the softest and most easily eroded rocks in Europe.

Figure 3.17. Distribution of sea cliffs in Europe, sites mentioned in the text are named

Nearly 22% of the coastline of Great Britain is cliffed, made up of both hard and soft rocks. Of this 4,063km or so some 2,705 km (14% of the total length of coast) is made up of cliffs greater than 20m in height. These predominate on the north and west coasts where hard rocks occur. By contrast, softer rocks derived from glacial material are more usual in the south and east.

Cliffs are extensive throughout the Mediterranean and Black Seas. Rocky 'karst' coasts with cliffs of varying height are present around the shorelines of eastern Spain, east of the Camargue in France, in Italy, southern Albania, Greece (Figure 3.18), and Turkey. In the Black Sea mountainous coasts with sheer mountains, cliffs, hills and narrow beaches make up one of the three main landscape types (soft shores with spits, barriers etc and coastal wetlands, saltmarshes etc. make up the other two). Eroding coasts of the Ukraine coast represent only 3% of its total length with only short sections on the southern coast of the Crimea resistant to erosion (Shuisky 1995).

These occur on the southern and southwestern coasts, the Caucasus and the southern and western Crimea. Limestone cliffs 250m high occur at a few locations, notably southern Dobrogea (northeast Bulgaria/southern Romania), whilst more fragile Pleistocene loess cliffs deposited under the influence of wind are present at several locations around the Black Sea coast.

Figure 3.18. Limestone cliffs on the island of Kephalonia, Greece

3.3 Nature conservation value

There are four key factors which influence the nature of the biological importance of sea cliffs:

1. Resistance of the geological structure - 'hard' verses 'soft';
2. Underlying rock type - calcareous versus acidic;
3. Geographical location - climate variability and incidence of salt spray;
4. and human use.

Each of these elements is reflected in the nature of the vegetation and its associated animal life. 'Hard' acidic rocks, e.g. Granite, exposed to salt spray support maritime vegetation (grassland and heathland) with rare plants. Near vertical slopes have nesting seabird colonies, especially in the north. 'Hard' calcareous rock cliffs, e.g. Limestone, are also subject to salt spray and may have maritime grassland with rare plants including many endemics. Nesting seabird colonies occur less frequently here, though breeding falcons and other birds of prey including Eleonora's falcon (*Falco eleonora*) in the Mediterranean, can be present. 'Softer' limestone rocks, e.g. Chalk, can support calcareous grassland and may include a number of 'lime-loving' rare plants. 'Soft' rocks which are neutral to acid, e.g. Pleistocene deposits, are less rich in species. Salt spray tends to influence only the lower slopes, if at all, as the eroded material often forms a beach at the base of the cliff. The open soils are especially important for invertebrates, ephemeral vegetation, wet flushes, scrub and woodland.

3.3.1 Vegetation

On stable 'hard' rock cliffs the plant communities tend to be represented by transitions which reflect the degree of exposure to salt spray and wind which in the most exposed location may over-ride the influence of the underlying rock type. Freedom from intensive human use also allows some rare and sensitive species to survive, including many which have been eliminated from the surrounding more intensively used inland areas. Included amongst these is a high incidence of endemism of some plant species intolerant of competition (notably *Limonium* spp.) which are found throughout Europe.

On 'soft' rock cliffs stability is of paramount importance; though both exposure to salt spray and wind may influence the type of vegetation which develops, this is much less obvious. Succession can occur on recent landslips when a period of stability ensues. The open nature of the disturbed surface is exploited by ephemeral plants and animal species, notably invertebrates of bare ground habitats. Given time, major cliff falls can become clothed in woodland as the slopes stabilise.

3.3.2 Plant community transitions and successions

Exposure to maritime forces is a key determinant of the type of vegetation which develops and is greatest on the most exposed shores where a long fetch generates high waves and swell and the prevailing winds help deliver salt spray to the cliff face and cliff tops. Climatic effects may also be significant and the amelioration of temperatures in temperate latitudes may help increase the geographical range and variation in the plant communities. When exposed to the full force of the effects of climate and salt-spray they become distinctively maritime. From a nature conservation point of view the development of maritime grassland or heath represents a rare and often diverse plant community. The more exposed, calcareous sites tend to be the biologically most diverse. The plant communities which develop on ledges or on the cliff tops composed of resistant rocks can also be relatively stable.

The vegetation contains plant (and animal) species which are mostly confined to the coastal fringe. Some species are, like those of saltmarsh, salt tolerant and include *Sedum rosea* (in the north) *Festuca rubra, Armeria maritima* and *Silene uniflora* (on the western fringes) and a variety of halophytes such as *Crithmum maritimum, Limonium binervosum* and *Salicornia* spp. (in the south including the Mediterranean). As the influence of saltspray declines upwards or inland from the breaking waves and out of reach of the maritime influence, especially on the cliff-tops, the grassland and heathland become more closely related to inland terrestrial types (Figure 3.19, redrawn from Mitchley & Malloch 1991).

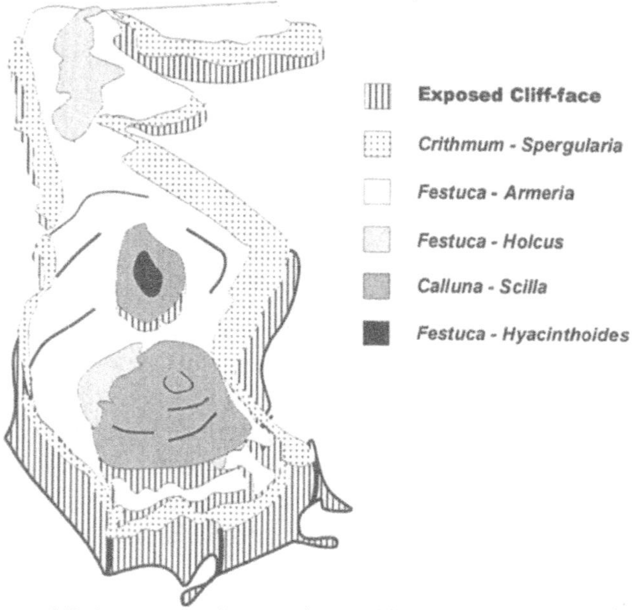

Figure 3.19. Sea cliff plant community zonation, an illustration from Gurnard's Head, west Cornwall, England.

Coastal climatic effects may extend inland for some distance, particularly at the top of exposed west-facing cliffs. The effect of this is often manifested by the shape of wind-pruned trees and shrubs which characterise some of the most exposed locations. In other areas oceanic effects may ameliorate the climate as on the west coast of Ireland where lichens and mosses of damp warm climates are found. Sheltered steep-sided coastal valleys and the lee slopes of coastal cliffs may support ancient woodland. The influence of salt spray recedes to the east, out of reach of the Atlantic gales. The megacliffs of Croatia are influenced by hurricane force storms, which can force salt spray up to 90m in the air (Lovric 1993a).

More mobile cliffs including those made of relatively soft geological strata may support open ephemeral communities on recent landslides. On very shallow soils certain species have adapted to avoid summer droughts by adopting a winter annual reproductive strategy; others have become xerophytic and possess thick succulent leaves and extensive root systems. The competitive ability of inland species is reduced by the high salinity and exposure allowing salt-tolerant species sensitive to competition to survive.

3.4 Plant communities - regional variation in Europe

Sea cliff vegetation is very varied. Three main types have been identified in the European Union Habitats and Species Directive: '**Vegetated sea cliffs of the Atlantic and Baltic coasts**'; '**Vegetated sea cliffs of the Mediterranean coasts with endemic *Limonium* spp.**'; '**Vegetated sea cliffs with endemic flora of Macaronesian coasts**' (European Commission 1999c). The first two of these are considered in more detail below.

3.4.1 'Hard' rocks of the north and west

Vegetation of 'hard' rock cliffs in the north exposed to the extremes of the gales of the north Atlantic have some of the best and most extensive examples of maritime rock-crevice and cliff-ledge community on the coast. Communities dominated by *Sedum rosea* and *Ligusticum scoticum* are found on many of the vertical salt spray drenched cliffs. On inaccessible cliffs away from grazing pressures *Silene uniflora* and *Armeria maritima* are also frequent. In some of the richer botanical areas Arctic-alpine plants such as *Saxifraga oppositifolia* and *Silene acaulis* occur.

Maritime cliff and cliff-top grassland occurs where grazing pressure is heavy and here the incidence of true halophytes and hence maritime influence is less obvious. The most common grassland community is cliff-top sub-maritime grassland with *Festuca rubra* and *Plantago maritima*. This is especially well developed on the islands of Orkney and Shetland in the United Kingdom where sheep grazing occurs in a region of high rainfall. Maritime heath is often associated with this community on less steeply sloping sea cliffs which are exposed to northerly and northwesterly winds and may also support transitions from grassland to heathland which can be rich in species, including rarities such as *Primula scotica*. In the north a maritime form of *Empetrum nigrum* heath occurs on deep, free-draining mineral soils whilst on wetter soils *Erica tetralix* may occur.

On relatively sheltered coasts where the force of the waves is dissipated at the foot of the cliff, the zone of the truly maritime vegetation will be limited to a narrow fringe just above high water. On the exposed Atlantic coastlines, communities composed largely of salt tolerant plants occur on cliffs to a height of as much as 50m, as for example at Yesnaby in Orkney, and may stretch some distance inland. In extreme situations around the edges of some 'Geos' on cliff tops in Scotland saltmarsh communities with *Glaux maritima*, *Festuca rubra* and *Juncus gerardii* develops in response to the continual drowsing by sea water.

Many cliffs (particularly those which are east-facing) are only marginally exposed to the onslaught of the full power of the breaking waves and salt spray. Here, either because of the orientation of the coast, the absence of a rocky foreshore against which waves can break, or because the cliffs themselves absorb the wave energy and erode, vegetation more typical of inland types occurs. On the east coast of Scotland this often includes dry and wet heath. In more sheltered spots, *Calluna vulgaris* heathland or tall grassland with bluebell *Hyacinthoides non-scriptas*, usually seen as a woodland plant, may occur.

3.4.2 Calcareous to neutral 'hard' rocks on the Atlantic coasts of Britain, Ireland, France, Spain and Portugal.

Cliffs of Limestone and other basic rocks are widely dispersed, but localised around the coasts of England and Wales and at locations in Northern France, Spain and Portugal. Although truly maritime communities occur, they are usually restricted to a narrow band at the base of the cliff. On the rock platforms characteristic maritime communities with *Armeria maritima*, *Crithmum maritimum* (Figure 3.20) and *Plantago coronopus* occur. They may also include the rare species *Parapholis incurva* and *Limonium recurvum*. In limestone areas the cliffs may provide important refuges for species, such as *Ophrys sphegodes*, typically found in inland grasslands, which have increasingly been destroyed by agricultural intensification. In some areas cliffs provide the most spectacular natural rock gardens such as Cape St. Vincent in Portugal, where the endemic *Cistus palhinhae* is one of the many rare species present.

On relatively sheltered, dry, calcareous cliffs *Brassica oleracea* is found on crumbling edges and sloping ledges. This rare species is in Britain found in association with other rare species such as *Silene nutans*. It also occurs at Etretat where chalk cliffs rise to 120m and grade into grassland and scrub woodland with *Fagus sylvatica* on uncultivated cliff tops (Géhu & Géhu-Frank 1993). Rich plant communities are also found on other basic rocks. The Lizard, Cornwall is one of the most important examples and the complex geology is matched by the rare and diverse nature of the plants and plant communities found there (Malloch 1971, Hopkins 1983). These areas also include transitions to heathland, normally with *Calluna vulgaris* and *Erica cinerea*, though in addition the Lizard has *Erica vagans*, a species rare in southern England.

Neutral to acidic substrata such as Granite and other resistant rocks are common along the more southerly European Atlantic west coast. Again exposure to salt spray results in maritime communities occurring at the most exposed sections of cliff. The transition is to maritime grassland with a high incidence of *Festuca rubra* and *Calluna vulgaris* heath often with a fringe of *Ulex gallii.* The best examples are found on the west coast of Wales, in Brittany, northern Spain and Portugal. In these last areas typically the cliffs support *Crithmum maritimum, Spergularia rupicola, Limonium binervosum* and *Asplenium marinum.* Out of reach of both salt spray and grazing animals, heavily wind-pruned *Ligusticum vulgare* and *Prunus spinosa* scrub may develop. *Erica vagans* and *Ulex gallii* are more common on the cliffs of Spain and Portugal.

Figure 3.20. Crithmum maritimum, common at or near the shore on rocky/cliffed coasts

Woodland is relatively restricted on all but the most stable cliffs, though inaccessible and sheltered sites may support important examples. On the island of Rügen on the German Baltic coast, for example, limestone cliffs and moraines (up to 120m) support communities ranging from open ground with *Tussilago farfara* community, to *Hippophae rhamnoides* scrub and *Fagus sylvatica* woodland. Here beneath the open canopy there are a large number of orchids including the rare *Cypripedium calceolus* (Hundt 1993).

3.4.3 Vegetation of the 'soft' rock cliffs of the southern North Sea, Channel coast and Baltic Sea

Land slip is common in sand and clay deposits in the south and east of England, Denmark and the southern coast of the Baltic. Where continuous erosion results in the formation of high sloping unstable cliffs, some of the least maritime of the coastal cliff vegetation is found. Plant communities can range from sparse ephemeral vegetation where movement is frequent to scrub and even woodland on some ledges where massive landslides occur and subsequently remain stable for tens or even hundreds of years. The landslip between Axmouth and Lyme Regis on the south coast of England on Christmas Day, 1839, is a famous example. Since then the undercliff has remained stable and now has mature woodland, open grassland and flushed areas rich in species and important on a national and European scale for its wildlife interest. Within the Wolinski National Park in Poland (Figure 3.21), steep unstable coastal Pleistocene cliffs support vegetation similar to that found on the more unstable areas at Rügen. At both sites and elsewhere on highly mobile cliffs, instability is important for the maintenance of open conditions necessary for the survival of species intolerant of competition.

Figure 3.21. Steep eroding glacial cliffs, Wolinski National Park, Polish, Baltic sea coast

3.4.4 Cliffs of the Mediterranean and Black Sea coasts

Unlike the corresponding cliffs in the north and west, the incidence of exposure and hence the maritime nature of cliffs in the Mediterranean are much reduced. This makes assessing their special importance, when compared to the rest, less clear cut. However the importance of the megacliffs of Croatia has already been alluded to. The description given by Lovric (1993a) for the Velebit coast provides a vivid picture of a coastal habitat of exceptional biological importance. The habitats range from emerging rocks with a calcifying algal community to sea water-drenched hyper-saline communities, including *Arthrocnemum glaucum* and *Atriplex portulacoides*, to exposed windswept cliff tops from 460-1,200m high. Interspersed in more sheltered areas are woodlands and scrub.

There is little in the literature concerning these habitats though reference is made to similar cliffs in Albania (Lovric 1993b), southwest Turkey (Lovric & Uslu 1993a) and along the western shore of the Black Sea, especially at Kiliakra (Lovric & Uslu 1993b). These areas should probably be considered alongside the Atlantic woodlands of the northwest and the laurel forests of the Canary islands for their conservation importance. All of these are in one degree or another dependent on their proximity to moisture-laden air derived from onshore winds.

3.5 Invertebrates

Sea cliffs are often very important habitats for a range of invertebrates. In Great Britain hard rock cliffs and cliff ledges support good assemblages of species, including specialists associated with carrion, such as *Trox perlatus* (an extremely rare beetle) and guano from sea bird colonies (Kirby 1992). Calcareous cliffs tend to have the most diverse fauna though all cliffs, both acid and calcareous, may have important species of wet seepages. South - facing slopes have richer assemblages of thermophilic species such as aculeate Hymenoptera. Some other species are specifically associated with cliff and cliff tops because of the presence of plants such as *Armeria maritima*, the sole food plant for thrift clearwing (*Bembecia muscaeformis*) whose larvae feed on its roots. Lovric (1993a, b) describes invertebrates on the cliffs of the Croatian and Albanian coasts (Table 3.2).

Table 3.2 Some rare invertebrates of Mediterranean coastal cliffs

Habitat	Species
Megacliffs of the northeast Adriatic (lower slopes only)	
Exposed up to 3 m (hurricane waves & winter ice)	*Mihovilia adriatica* (a gastropod)
Exposed 3 - 30 m (stormy swash)	*Ligia italica* (an isopod)
Sheltered ravines up to 4 m (stormy surf)	*Talytrys platycheles* (a crustacean) & *Desidiopsis* (a spider)
Loamy (Palaeozoic) coastal slopes, Istra Peninsula	
Maritime ledges (20 m landwards)	*Delima decipiens* (a snail)
Submaritime scrub (up to 150 m inland)	*Eobania vermiculata* & *Monacha olivieri* (snails)
Salt spray semi-evergreen pseudomaquis	*Helix cincta* & *Milax gagates* (snails)

'Soft' rock cliffs are also important but have a very different fauna. The bare ground associated with the frequent slippages is often rich in species including several, which in Britain at least, are found in no other habitats. The important habitat features are listed in the Table 3.3 below together with an indication of some interesting species taken from Kirby (1992).

Table 3.3 Sub-habitats on soft rock cliffs important for invertebrates in Great Britain

Habitat	Species
Bare ground subject to frequent slippage	Solitary bees and wasps (e.g. *Eucera longicornis*)
Steep slopes with open vegetation	Weevils (e.g. *Sitona gemellatus*)
Cracks and crevices	Ground beetles (e.g. *Nebria livida*)
Wetland (seepage zones)	Shorebugs (e.g. *Saldula arenicola*)

3.6 Human activities and conservation

Sea cliffs represent a substantial landscape resource. Vertical or near vertical cliffs although generally inaccessible, are not immune from human intervention. The vegetation of some of the steepest slopes has been influenced both directly and indirectly by human use. On the cliff top and more gentle slopes impacts are much greater. Activities affecting these areas can be broken down broadly into two types: those which cause direct destruction of the habitat and those which modify it, perhaps creating other types of interest. Infrastructure development, including hotels providing a 'room with a sea view', agricultural intensification and recreation have all taken their toll and in some areas losses of semi-natural habitat have been considerable. As is shown below, these losses are exacerbated by other land-use changes such as grazing management. A summary of the main management issues influencing the conservation of sea cliff and cliff top vegetation is given in Table 3.4. The following discussion is based largely on the situation in Europe; a wider view of cliff ecology and the factors influencing pattern and process is provided by Larson et al.(1999).

Table 3.4 Key management issues affecting sea cliffs in Europe

HABITAT LOSS:	Horse riding;
Agricultural intensification, including reseeding and conversion to arable land;	Fires;
Housing and tourist development;	Disturbance including rock climbing.
Building roads and other infrastructure including industrial development;	ENGINEERING, & SEDIMENT MOVEMENT:
Aggregate extraction from coastal quarries.	Coastal protection structures;
GRAZING -RELATED MANAGEMENT:	Remedial engineering, including drainage and slope stabilisation;
Overgrazing/undergrazing;	Climate and sea level change;
Establishing grazing regimes;	The importance of storms.
Burning, as a management tool;	OTHER MANAGEMENT ISSUES:
Pteridium aquilinum control.	Invasive/alien species;
TOURISM & RECREATION:	Pipe-laying, including gas and oil;
Car parks;	Acid deposition;
Trampling/erosion control;	Eutrophication & pollution;
	The importance of desiccation.

3.6.1 Habitat loss - agricultural

The amelioration of the climate, which often characterises the maritime zone in many temperate regions of northwest Europe, has helped create favourable conditions for agriculture, amongst other uses. The more stable cliffs and accessible cliff slopes have been used historically as grazing land, including over-wintering cattle or sheep in northwest Europe or summer grazing in the Mediterranean ("transhumance", Blondel & Aronson 1999). These activities have modified the vegetation, helping to create species-rich grassland and heathland, preventing the development of scrub and woodland.

In recent years more intensive farming methods have replaced pastoral use of the land with the result that the rich vegetation has been lost directly through ploughing and planting of arable crops, or reseeding of pastures. This has destroyed some of the best examples of grassland and heath in areas such as Orkney, in the UK, where maritime heath has similarly been ploughed to 'improve' agricultural production. In southwest England and in the Pas de Calais region of northern France, rare examples of western heathland have also been lost on cliff tops. In some areas, particularly on the softer rocks of the east coast of England, agricultural land may reach right to the cliff edge, especially where erosion rates are high and cliffs readily collapse. The glacial (Pleistocene) cliffs on the Holderness coast, for example, which have retreated at an annual average rate of 1.8m since 1852

(Lee 1995), have virtually no semi-natural vegetation for this reason. Even in areas such as Beachy Head, where semi-natural cliff-top grassland remains, it is being squeezed into an ever narrower belt as the chalk cliffs recede. It seems certain that in time the combined effects of the two processes will result in the total destruction of some of the most important cliff-top grassland in southeast England.

Quantification of these changes is difficult because of the absence of good base-line surveys. However, in Great Britain comparison between the 1st Edition of the Ordnance Survey 6 inch (1:10,560) County Series maps of 1889 and the more recent 1:10,000 of 1963 have been used to provide some insight into the change in use of the coastal cliff tops of North Cornwall from Lands End to the Devon border (Bennett 1984). The two series of maps show that there has been an overall reduction of 21% in the area of 'rough pasture' or 'scrub' as defined on the 1st edition map. Accompanying these changes there appears to have been a decrease in grazing. Bennett (1984) concludes that: -"One of the main factors causing change to cliff-top habitat is the interaction of recreational use of the path by walkers with farmers. Farmers are reluctant to allow sheep and cattle to graze the cliff tops due to the possibility of dogs worrying the stock". These results provide only a measure of the scale of loss without implying losses to the quality of the vegetation.

3.6.2 Habitat loss - tourists, urbanisation and engineering structures

Attractive scenery, fresh air, warm climate and the alleged healing properties of sea water attracted much development to the south coast of England early in the last century. Even in less climatically favoured areas such as the south coast of England, housing developments when accompanied by caravan sites, roads, car parks and other tourist attractions can result in a major loss of habitat and impact on the landscape. A room with a 'sea view' was all important and cliff tops offered particularly fine locations. Throughout the rest of Europe, hotels and villas on the Atlantic coast of France, northern Spain and Portugal, and on the limestone cliffs and slopes around the Mediterranean are extensive and in places densely populated.

'Soft' rock cliffs have an entirely different set of management issues notably associated with the desire to prevent landslides and erosion in order to protect infrastructure development, again perched on the cliff top, often in inappropriate locations. The massive growth of 'Time-share' apartments has resulted in an increase in this type of development. The result of this has been to increase the desire to stabilise the cliff slopes and/or prevent erosion of the base of the cliff through marine action. On cliffs with open soils and seepage zones where there are specialist communities, this can result in an overall loss of conservation interest as coarse grassland, scrub and woodland develop. Although not directly related to the conservation issues discussed here this may also have an adverse impact on geological sites where periodic erosion exposes their important features.

3.7 Grazing

Grazing by domestic stock has probably been taking place in some areas, such as the cliffs of the Lizard in Cornwall, since prehistoric times (Hopkins 1983). It has certainly had an important part to play in the development of sea-cliff and cliff-top grassland and heath on all but the steepest slopes in Great Britain (Mitchley & Malloch 1991) and probably elsewhere in Europe. Without it, except in the most exposed locations where wind, salt spray and exposure restrict the more vigorous plants, woodland and/or scrub becomes the prevalent vegetation. Its importance can be gauged by reference to the management needs of the many kilometres of coastal cliff and cliff-top habitat in England and Wales owned by the National Trust (a non-governmental organisation). No fewer than 13 Red Data Book plant species (some of the rarest species in Great Britain) occur in maritime heath and grassland on these sites and where grazing is a significant factor in their survival (Hearn 1995).

In recent years the conservation value of these open areas has been adversely affected by the abandonment of grazing. The reduction in regular stock grazing, combined with the loss of the rabbit population in the early 1950s, resulted in growth of coarse grasses, bracken and scrub at the expense of species-rich grassland and heath. Management effort in many coastal cliff nature reserves affected by these problems has concentrated on the control of scrub, including the introduction or re-introduction of grazing by domestic stock (Mitchley & Malloch 1991).

Assessment of which grazing regime is appropriate is difficult due to changes in the nature of the vegetation, climate and the behaviour of the animals. Some guidelines are given below on appropriate grazing regimes for a range of coastal grasslands based on the detailed study of some individual cases in the UK (Table 3.5 from Mitchley & Malloch 1991). At higher stock levels the nature conservation interest may be low or non-existent. Lower stock levels may be insufficient to remove or control coarse grasses and scrub.

Table 3.5 Indicative stocking rates of sea cliff vegetation from case studies in the UK

Habitat Vegetation	Grazing category	Grazing regime	Comments
Improved grassland	Very high	15 sheep $ha^{-1}yr^{-1}$.	Assuming inputs of artificial fertiliser
		7.5 sheep $ha^{-1}yr^{-1}$	No artificial fertiliser
Semi-natural pasture	High	4.0 sheep $ha^{-1}yr^{-1}$	
Calcicolous grassland	Medium	2.5 sheep $ha^{-1}yr^{-1}$	
Calcifuge grassland	Medium	2.0 sheep $ha^{-1}yr^{-1}$	
Maritime grassland	Medium	2.0 sheep $ha^{-1}yr^{-1}$	
Limestone heath		0.5 sheep $ha^{-1}yr^{-1}$	
Tall heath (lowland)	Low	0.5 sheep $ha^{-1}yr^{-1}$	
Maritime heath	Low	0.5 sheep $ha^{-1}yr^{-1}$	
Wet heath / bog	Very low or none		
Scrub	Very low or none		Depending on management aims

Stock numbers must be balanced between levels of grazing which can convert heathland to grassland (overgrazing) and undergrazing which allows more invasive species such as bracken and gorse to enter. Because of the low productivity of heathland, a detailed review of lowland heath management (Gimingham 1992) concluded that grazing as low as 0.5 sheep per ha per year, or less might be appropriate. The balance struck must depend on a good knowledge of the importance of the existing nature conservation interest. In most cases local knowledge will provide this, though in effect the decision will be taken on the basis of a variety of factors. Mitchley and Malloch (1991) give a series of steps to be taken in making a decision for grasslands but they are equally appropriate for coastal heathlands (and indeed other coastal habitats such as sand dunes and saltmarshes) where grazing management forms an important element in the development and survival of the nature conservation interest. These have been modified to give the following suggested scheme:

- Step 1 What is the nature of the existing nature conservation interest?
- Step 2 Is grazing required to maintain the interest?
- Step 3 Is the current grazing regime adequate to maintain the interest?
- Step 4 If damage appears to be occurring consider changing the regime - but slowly!
- Step 5 Review past management, adjust stocking rates and timing; monitor effects.
- Step 6 Assess efficacy of management - repeat cycle steps 3 - 6.

The steps suggested above are deliberately vague as there is no guaranteed prescription for managing vegetation by using grazing animals.

However, a good knowledge of the site and the prevailing climatic and edaphic conditions are all important.

In areas of intensive stock rearing, adjacent to cliffs, eutrophication can also be a local, but important issue. Application of artificial fertiliser will reduce the overall species richness of the vegetation, though this is seldom applied directly to the cliff slopes. In some areas adjacent cliff slopes may provide important supplementary areas for less intensive use, for example as seasonal grazing. This can result in the transfer of excess nitrogen from the intensively managed areas to the cliff where animals may shelter.

Little is known about the impact elsewhere in Europe, though grazing is important to the maintenance of cliff grassland and scrub on the Danish Island of Samsø. Here the animals create terraces and help favour species which tolerate and benefit from grazing (Jensen 1993). The knock-on effects of changes in patterns of grazing for associated animals is also important. In Croatia loss of traditional sheep grazing reduces available food for griffon vulture (*Gyps fulvus*) and threatens its survival, especially on Cres Island in the northern Adriatic (Bakran-Petricioli et al. 1996). The cases of two butterflies and the chough (*Pyrrhocorax pyrrhocorax*) are discussed below.

3.7.1 Butterfly conservation - a cautionary tale

Manage the vegetation and the invertebrates will look after themselves! To some extent a suitably managed grassland or heath will include species, notably invertebrates, with good populations of rare and local species. Many of these may be specific to the cliff habitat, others find refuge from the destruction of inland habitats where they may have formerly thrived. Whatever their origins, a few species have very special needs not always catered for by the general approaches to management. This is particularly true for species with a life-cycle which is complete in one season.

Butterflies, particularly those with a southern distribution in Europe, are found on cliff habitats further north than the centres of their distribution because of the microclimate conditions of south-facing, often hot dry slopes. The presence of mature vegetation with a good structure and mosaic of species suitable for the caterpillars and nectar for adults result in good populations of a wide variety of species. However, minor changes in management can have a significant, if unforeseen, effect on a single species. The story of the large blue butterfly exemplifies the complex interactions which suggest that quite minor changes in grazing regimes can have devastating effects on some species, especially those at the edge of their geographical range.

The large blue (*Maculinia arion*) is a butterfly often found on the European continent in large numbers where it is quite tolerant of different heights in vegetation. The population in Great Britain had declined from nearly ninety sites in the 1700s to thirty in 1950s and four in the 1960s. By 1979 the final colony, in South Devon had become extinct. The reason for the decline was not fully understood. Though habitat loss from ploughing of open calcareous grassland, afforestation, building and quarrying had all taken their toll. At several sites the extant vegetation seemed suitable, with abundant *Thymus polytrichus*, the main food plant of the young caterpillars.

Despite major efforts in the 1960s to rehabilitate their habitats, the few surviving colonies failed. One of the principal reasons appeared to be the growth of rank grassland as a result of decrease in grazing brought about by the reduction in livestock grazing and the demise of the rabbit population following myxomatosis in the 1950s. At the time the fact that the species relied on the presence of the red ant (*Myrmica sabuleti*) to carry the butterfly larvae (developed from eggs laid on *Thymus polytrichus*) to underground nests, where they feed on ant grubs, was not fully understood. This species of ant requires close-cropped grassland and even a small relaxation in grazing pressure can result in its replacement with an unsuitable species *Myrmica scabrinodis*. Although *Thymus polytrichus* might be present in abundance, without the host ant, the population died out (Thomas 1989).

Another example derives from northeast England. Here a sub-species of northern brown argus (*Aricia artaxerxes*), 'Durham' or 'Castle Eden' argus (*Aricia artaxerxes salmacis*) thrives on Magnesian Limestone grasslands along the coast. Unlike the large blue, which not only has a special association with an individual ant species and requires short-cropped grassland (<3cm, optimal 1cm) the Durham argus needs an intermediate, 6-10cm turf, for its survival.. For this species there is no specific association proved with ants and winter grazing is the preferred. Heavy spring and summer grazing can cause population crashes (Ellis 1997).

3.7.2 Sea cliffs and the chough

The chough is a bird of mountains and coastal cliffs nesting in crevices and caves. Although widespread in Europe it is rare and declining across its range. In the United Kingdom and Ireland it is restricted to coastal cliffs where it feeds on invertebrates in areas with short-grazed turf (Figure 3.22). Persecution in the 19th Century and more recent changes in agricultural practice are often cited as the main causes of decline. Agricultural effects include loss of cliff top grassland through intensification and a reduction in sheep, cattle and rabbit grazing allowing coarse scrub to develop (Bignal & Curtis 1989).

Figure 3.22. Grazing on sea cliffs helps create short grassland turf, essential for a variety of rare plants and animals such as the chough

In addition to these effects recent research on the species suggests that learnt social behavioural characteristics may also be important. It was found that co-operation with kin and the flock helped survival during periods when prey was limited. A relatively predicable environment supporting a large population of individuals is necessary for this social organisation to develop. The mosaic of habitats necessary to facilitate this must be on a scale large enough to allow the chough to take advantage of suitable conditions in good years to replenish the population. When considering changes to land management for conservation purposes it is important to look at the long term implications rather than just reacting to short-term trends in population numbers (Bignal et al. 1997). This example also illustrates an equally important point, that management of individual species may require more than the traditional habitat protection and management. An understanding of the behaviour of species and the intra-relationships between individuals, an area hitherto not considered in detail, may also be needed.

3.7.3 Restoration

Rehabilitation of undergrazed grassland or heath which has become overgrown with moribund heather, bracken and/or scrub is one of the most frequently encountered management issues on coastal cliffs of northwest

Europe. Re-introduction of grazing, burning, mowing and or mechanical control have all been tried with varying degrees of success. Extensive management has been undertaken in the UK, notably by the National Trust (Hearn 1995). Trials by English Nature, the statutory body for promoting conservation in England, have tested grazing regimes under its 'Wildlife Enhancement Scheme' on Magnesian Limestone including the coastal cliffs in County Durham and found stocking rates of 1-2 cattle or 4-10 sheep per ha are effective in restoring short grassland, 0.5 - 1.0 cattle or 2-5 sheep per ha are required for their maintenance. In this regime grazing is restricted from mid-June to November, allowing grassland plants to flower and set seed.

3.8 Recreation

With the increase in leisure activities the spectacular nature of the cliffed coastline of many parts of Europe has proved a powerful attraction. This in its turn has led to developments which in remote areas have detracted from the natural landscape and wildlife of the area. Visually intrusive caravan sites, car parks and other buildings on cliff tops are common. Around the coast of Great Britain, for example, approximately 75 caravan sites and 110 car parks are shown on or near cliff tops on the 1:50,000 Ordnance Survey maps. In addition to the physical destruction caused by the siting of these developments, the peripheral effects associated with informal refuse disposal, access points and footpaths can also adversely affect nature conservation interests.

Natural erosion rates can also be accelerated by the creation of paths on unstable slopes. Gullying of cliff tops and the transport of material down the slope are accompanied in extreme situations, such as at Hengistbury Head near Bournemouth in the county of Dorset, by loss of cliff-top and cliff-face vegetation (May 1977). Other impacts may be visually intrusive but without the same damaging effects. On the more stable cliffs, such as Kynance Cove in southern England, the passage of large numbers of people (more than 175,000 per annum) from the car park to the beach has created a marked track (Goldsmith 1977). Away from the relatively narrow track, species-rich grassland survives in an open sheep-grazed sward.

In southwest Portugal the Sagres Peninsula is also heavily trampled. The plant communities are fragile, partly due to the shallow nature of the soils in which they exist. Removal of the soil has led to the invasion of weeds of which *Carpobrotus edulis* is the most prominent. Although the nearby Cape St Vincent is less heavily impacted, both sites are of such conservation importance that they are deserving of special protection from overuse by visitors attracted by the landscape and historical interest rather than the flora and fauna (Woodell 1989).

Trampling effects have been described according to their damage to vegetation (Liddle 1975, Packham & Willis 1997 p. 261). Burden & Randerson (1972) also suggest upper limits when trampling acts in a destructive manner. However, this work also indicates that trampling can be a positive force in maintaining open grassland communities, especially where more sensitive and rarer species are being lost. Thus in some instances human pressure might be used to control coarse grasses and in limited areas combat the effects of under-grazing.

3.9 Other management issues

Listed in the issues affecting coastal cliffs are a number of other activities, all of which potentially pose a threat to the survival of cliff and cliff-top wildlife. Two of these are considered briefly below.

3.9.1 Tipping

Commonly tipping of old cars, and other rubbish on remote cliffs is visually intrusive though may have little impact on the conservation interest. On the Durham coast, northeast England, however, there has been a long history of the tipping of colliery spoil, both offshore and on the cliff top. In addition to the direct destruction of the cliff vegetation, offshore disposal has raised beach levels and reduced the number of occasions on which the sea can attack the base of the limestone cliffs. This has helped create greater stability in the overlying boulder clay, which has aided the development of species-rich grasslands, more typically found on inland Magnesian Limestone (Doody 1981).

Paradoxically, this grassland is threatened by the cessation of tipping which occurred several years ago and attempts to clean the beaches and attract greater recreational use. Studies undertaken between 1991-1993 show that the beach is steepening and narrowing. When taken together with a lowering of the beach due to subsidence, the beach acts in a similar way to one where sea level rise is occurring. This results in the base of the cliffs being exposed to more frequent wave attack, in some instances for the first time (Humphries 1996). The long-term effects could result in the demise of the more mature species-rich grassland and with it colonies of the rare sub-species of the northern brown argus (the Durham argus) particularly those developed on the less stable substrates as suggested by Doody (1981) and Cooke & Gray (1984). A proposal to tidy up the whole of the coast with money from a successful millennium bid (East Durham Task Force 1995) may further aggravate these problems.

3.9.2 Invasive species

Invasive species are a small but increasing threat to cliff vegetation in western Europe. Species such as *Carpobrotus edulis* and *Disphyma crassifolium*, plants from South Africa grow vigorously on sunny cliffs and can cause a problem for some native species. Both species can be seen cascading down cliff slopes from as far afield as northwest Scotland (*Carpobrotus edulis*) and southern Portugal and Spain. The thick, matted vegetation of the former smothers all but the most vigorous species. As a consequence it is particularly damaging to cliff vegetation where it can quickly replace rare endemics which survive here because of the freedom from competition from other more vigorous species. Some rare clovers such as *Trifolium incarnatum molinerii* and *T. strictum* growing on the cliffs of the Lizard Peninsula, Cornwall have been adversely affected in this way (Hopkins 1979).

Chemical control of these and related species is possible, but the removal of the matted litter layer is effective only if done manually, a not inconsiderable task. *Disphyma crassifolium* does not produce such a deep litter but the roots penetrate into deep rock crevices and is hence also difficult to remove. In many areas the only management option is to leave the plant to its own devices. Some, other species pose little or no threat and may even enhance the visual beauty of an area, as in Madeira (Figure 3.23).

3.10 Conservation and coast protection

The above discussion shows that many areas require management action to sustain their nature conservation interest. Even seemingly inaccessible cliffs may be found to have been managed in the past. Beast Cliff in North Yorkshire in England, for example, is a plateau lying 180m below a steep apparently inaccessible cliff with a further steep drop to the sea. Despite its apparent inaccessibility, in the past animals were traditionally let down on ropes to graze the vegetation in the summer.

Some of the most important sea cliffs and their vegetation described in this chapter are those with the greatest exposure to salt spray along the Atlantic fringe of northwest Europe. Other exposed, though more restricted, sites exist in the Mediterranean and Black Seas. They are often composed of hard rock and located on west-facing slopes in areas where local topography funnels salt water over the cliff to saturate the soils. In the most exposed locations the transitions from saline communities through to maritime grassland and heath may require little or no management as soil creep and exposure help to keep the vegetation open and free from scrub and woodland. By contrast where human activity has modified the nature of the vegetation, for example by managing the cliff slopes by grazing, cutting of furze etc. and these activities have ceased, reintroduction of active management may be required.

Figure 3.23. Kniphofia uvaria (red-hot poker) naturalised on coastal cliffs in Madeira

At the other end of the spectrum the softer clay cliffs, if left to themselves will continue to respond to the natural forces causing slippage and erosion, helping to create open, rich plant and invertebrate habitat. Where infrastructure development has taken place constraints on the natural movement of the cliffs may be deemed necessary to protect property. In addition to the direct destruction of the cliff top habitat, the erection of solid sea defence structures is often destructive, and obliterates some of the most maritime communities at the base of the cliff. They also help to 'lock-up' sediments normally available for tomorrow's beach, saltmarsh or sand dune. These 'knock-on' effects are important for conservation of coastal wildlife and may have economic implications.

3.10.1 Economics and conservation

The more unstable the cliff and the more densely populated the hinterland, the greater the desire to prevent their erosion. This has resulted in various protective measures including groynes, sea walls (e.g. revetments) and strengthening and drainage of slopes, all designed to prevent cliff slippage and/or erosion by the sea. This in turn can result in less frequent movement and loss of open unstable forms of vegetation and its associated invertebrate fauna. The impact both on the cliff itself, including its geological value and the implication for adjacent areas, are key issues. Taking the conservation interest first, many of the more unusual plant and animal communities of unstable cliffs require an open slope with sparse and/or ephemeral vegetation for their survival. Stability results in growth of coarse grassland and scrub at their expense.

Having established expensive structures close to the sea it is understandable that those living or investing in them will seek to protect them if they are threatened by erosion. However, in addition to the potential loss of conservation interest as the slopes become more stable, the nature of the cliff and its location within the wider sedimentary system are also important considerations. On the coast of the Algarve the rapid expansion of the tourist development has occurred on the Pleistocene cliffs (5 - 30m high). Erosion rates of 0.7m per year at Forte Novo increased to 1.4m following the building of a jetty at Vilamoura.

A similar doubling of the erosion rate (from 0.6 - 1.0m to 2.0m) also occurred at the prestigious nearby Timeshare complex when rip-rap was placed on the shore to protect a swimming pool (Correia et al. 1996). In this area not only have the extensive pine forests which covered the cliff been destroyed, but the survival of the development itself is threatened because of the continuing erosion of the cliffs by the action of rain and removal of the beach by waves (Figure 3.24). A further consideration is the extent to which eroded material is available and provides sediment for other structures such as beaches and dunes, themselves natural coast protection features. Without them there is often a need to continue protection further along the coast.

Perhaps the lesson from this lies in better understanding by those who are responsible for siting developments of the economic (and conservation) implications of building on naturally unstable cliffs. Another case, in northern Portugal, highlights the issues. Rapid growth in tourist urbanisation south of the Cávado River coupled with increased erosion of the shoreline elicited the classic engineering response. A series of revetments and groynes were built that temporarily protected the buildings. However, the impact of the interruption of long-shore drift to the south had disastrous consequences, threatening the small village of Pedrinhas (Granja & Soares de Carvalho

1991). In the longer term perhaps retreating from a rapidly eroding coast is both environmentally and economically more sustainable than fighting to stay put. Better still avoid building on unstable cliffs in the first place.

Figure 3.24 Eroding cliffs and 'swimming pool', near Vilamoura, the Algarve, Portugal

3.10.2 Stay back or set back?

The examples quoted above beg the question as to the suitability of some coastal areas to sustain habitation. Along the Holderness coast in eastern England many settlements have fallen into the sea since Roman Times and an area of some 83 square miles of land is estimated to have been lost (Steers 1969a). Prevention of erosion here, as in other major soft coastal areas, has not taken place due to the scale of costs. However, elsewhere this is not always the case. There are important lessons here for the engineer as gabions or similar structures are expensive and not always successful (Figure 3.25). In these situations the best protection for a cliff is a good beach. Staying back from the hazard zone and moving inland may also provide a less costly and more long-lasting solution. Allowing more natural processes to work may be the best option from an economic point of view and can even have conservation benefits.

Figure 3.25. Engineering solutions can be costly and ineffective; gabions on the Polish coast

4. SEABIRDS, SEA CLIFFS & ISLANDS

4.1 Introduction

The previous chapter dealt with the conservation of sea cliffs and sea cliff vegetation. This chapter briefly considers the conservation of a variety of mobile species (notably seabirds) inhabiting sea cliffs and islands. Historically, many island bird and mammal populations have provided a source of food for indigenous populations. Larger flightless birds such as geese were particularly vulnerable to human predation and in Hawaii Polynesian settlers are estimated to have exterminated 39 species (including 7 species of geese) over the last 1,000 years. A further 14 have been lost in the last 200 years as European colonisers continued the extermination of native species (Schreiber et al. 1987). The introduction of ground predators, notably black rat (*Rattus rattus*) and in Europe American mink (*Mustela vison*), caused further losses and in many areas today populations are considerably reduced or extinct.

No attempt is made to deal with the world-wide problem of extinction highlighted above. However, the chapter does look more closely at management issues as they impact on the survival of seabirds and other coastal species of cliffs and islands. The implications of oil pollution, over-fishing and other activities, and their cumulative impact on bird populations, are also considered.

4.2 Seabird locations and habitat

Colonies of cliff-nesting seabirds are a significant interest on many sea cliffs and islands throughout the World. In both the northern and southern hemispheres particularly large breeding colonies are attracted by the presence of a plentiful and accessible food resource in the rich waters of the Arctic and Antarctic oceans. At some individual locations concentrations can be measured in millions of individuals. This chapter is confined to a discussion of species inhabiting the cliffs and islands surrounding the coast of the northeast Atlantic and in the Mediterranean. As many seabirds are ground-nesting they are particularly vulnerable to predation. Because of this, isolation and inaccessible nesting sites are required, in addition to being close to abundant food resources. Rocky shores, steep high cliffs and exposed short-turfed grasslands are the principal nesting habitat. Where conditions are favourable large colonies occur around the margins of the sea. The concentration near the rich waters of the north Atlantic is illustrated by puffin (*Fratercula arctica*) which is virtually absent as a breeding species further south (Figure 4.26).

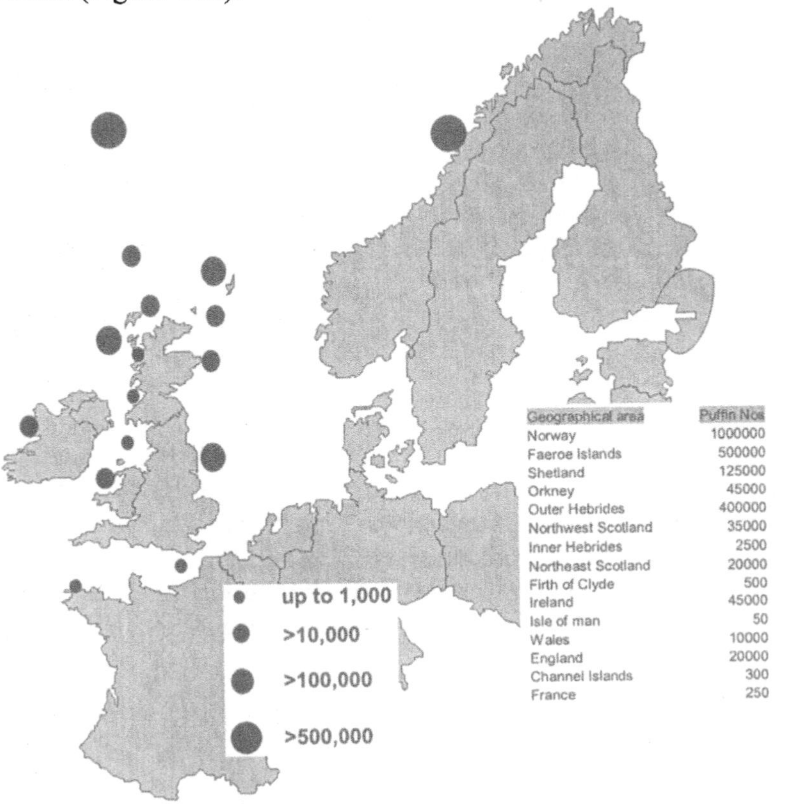

Figure 4.26. Puffin colonies (numbers of breeding pairs) in northwest Europe (Harris 1984)

In northern and western Europe, including Scandinavia, seabird nesting habitats include: ledges on steep cliffs e.g. guillemot (*Uria aalge*), razorbill (*Alca torda*) and kittiwake (*Rissa trydactyla*); amongst rocks black guillemot (*Cepphus grylle*), or in burrows on cliff-top and island grassland shearwaters and puffins. The northern, cliff-nesting seabirds decline in numbers towards the south where different species appear, including shearwaters and petrels. These include the very rare petrels *Pterodroma madeira* and *P. feae*, known only from the island of Madeira. Other species such as the gulls Audouin's gull (*Larus audouinii*) and Mediterranean gull (*Larus melanocephalus*) occur in significant numbers in the Mediterranean and Black seas.

4.3 Seabird conservation and human activities

The conservation of seabirds is affected by a variety of human activities which have a direct impact on individual birds, their nesting habitat and the abundance and accessibility of their food supply. Some of the habitat issues are dealt with in the previous chapter. These include agricultural intensification and grazing management, which may be particularly important on some islands. In addition to these habitat management issues, seabirds are directly exploited for food, vulnerable to the impact of external influences such as oil pollution, and affected by over-fishing. Disturbance to nesting birds even on the steepest, seemingly most inaccessible slopes may also occur where climbing takes place. Table 4.6 summarises these issues.

Table 4.6. Key management issues affecting seabird conservation

DEVELOPMENT:	Invasive plants.
Coastal development *.	**FISHERIES**
DIRECT EXPLOITATION:	Over-fishing small fish *;
"Harvesting" adults & young;	Reduced discards & offal *;
Egg collecting and hunting.	Seabird by catch e.g. gill nets
EFFECTS ON BREEDING COLONIES:	**INDIRECT EFFECTS:**
Introduction of predators *;	Oil pollution *;
Increase in native predators *;	Other pollution e.g. plastic;
Disturbance of nesting birds *;	Dredging *;
Recreation e.g. rock climbing;	Coastal erosion *;
Vegetation change *;	Climate change.

* Activities identified as having a potentially critical or high impact on seabird colonies (Tucker & Evans 1997)

4.3.1 Exploitation

Harvesting of both seabirds and their eggs for food was a major activity in the past even on the steepest most dangerous slopes. In many areas of the world it sustained isolated island and coastal communities. Although this activity is now much reduced it is still practised by some communities, especially those where other food resources are difficult to obtain. Today, for example, the puffin is a traditional food resource on the Faeroe Islands, fulmar petrel (*Fulmarus gracialis*) eggs are gathered from islands off the south coast of Iceland, short-tailed shearwater (*Puffinus tenuirostris*) are harvested from their burrows in Tasmania and glaucous gull (*Larus hyperboreus*) is an important food for Greenland Eskimos. The precise impact on the populations is unclear, but it is known that vast numbers of individuals were once taken. Some species appear to show a decline, apparently due to this exploitation, whilst others show no appreciable effect (Schreiber et al. 1987).

Although predation by explorers may have ceased, annual culls still take place in some areas. On the island of Sula Seigra off the west coast of Scotland, for example, gannets (*Sula bassana*) are taken for food. These are not required to sustain the human population but the activity is maintained as a traditional and jealously guarded right (Beatty 1992). There is little evidence that the limited amount of traditional seabird harvesting being carried out in the northern hemisphere is today having any direct impact on the viability of the exploited populations. Indeed it is not in the long term interests of indigenous people to over-exploit the resource. However because of the links with other effects such as over-fishing, oil pollution and the introduction of predators the precise situation is difficult to assess.

4.3.2 Seabirds and island predators

Islands (and sea cliffs) provide havens from natural predators and hence are the favoured location for many ground-nesting seabirds. Species such as gannet may choose some of the most inaccessible places to breed (Figure 4.27). Throughout the world predation by introduced species is an important issue affecting these populations and is one of the most serious conservation problems in Europe today (Tucker & Evans 1997). Zino's petrel is threatened with extinction on the island of Madeira from introduced rats and feral cats (Zino et al. 1996). Changes in the distribution in Iceland of black guillemot and a 70-80% loss of a population of Arctic tern (*Sterna paradisaea*) at one colony in Finland (Moors & Atkinson 1984) have been attributed to mink. On some islands in Estonia, red fox (*Vulpes vulpes*) and racoon dog (*Nyctereutes procynoides*) prey on island nesting birds (Ratas & Nilson 1997).

Direct loss is not the only important factor in determining the management response to introduced predators. Breeding success of short-tailed shearwater in Australia was studied at two sites, Cape Woolamai and Benison Island, Victoria, to investigate the role of foxes (introduced in 1860) in the decline of the species population. The results suggested that the presence of other prey species, mainly rabbits (*Oryctolagus cuniculus*) reduced the impact on the shearwater population. Comparison of this study with data from other Victorian ringing sites showed the relative unimportance of predation on all but Benison Island, where no other vertebrate prey were present (Norman 1971).

This work illustrates that the control of predators must take account of the inter-relationships with other potential prey species. Tucker & Evans (1997) suggest that one of the principal ecological targets for seabird conservation in Europe lies in eradicating "introduced predators...where feasible." Where feasible is an important qualification. The number of islands where this issue is prevalent is daunting in itself. It is also doubtful if complete eradication is possible on all but the smallest islands. For example in the Mediterranean it is suggested successful eradication of rats can be undertaken on islands of less than 30ha. For other larger species, cats and goats, islands up to 2,000ha can be cleared (Pérez 1997). Since control of predators, once established, is difficult, their introduction, accidental or otherwise, should be avoided.

Figure 4.27. Nesting Gannets on the west coast of Scotland

4.3.3 Oil pollution

Many seabirds nesting on cliffs or islands are influenced by activities taking place away from the breeding site. Most obvious are the effects of oil pollution and food availability. The former can have a major and highly visual impact on individual birds when large accidental oil spills occur. Under these circumstances many birds and other animals can be affected and each incident can elicit intense media interest. During the infamous Torrey Canyon disaster off the coast of Cornwall in 1967 considerable expense was incurred spraying detergent which was relatively ineffective and harmful to marine and shore life (Smith 1968). Less obvious but possibly more significant to wildlife in the longer term are the numerous accidental and deliberate spillages. These may occur when cleaning oil tanks at sea, from land-based sources or from oil production platforms. Taken together, the cumulative effects are difficult to assess though it has been claimed most seabird mortality occurs as a result of this chronic pollution (Clark 1984).

Analysis of the recent 'Braer' incident, off the coast of Shetland in northern Scotland, suggests that coastal habitats and seabird populations may be more resilient to major incidents than first impressions suggested (Ritchie 1995). Though in the case of the 'Braer', stormy weather and light oil reduced the adverse impact on the environment and its wildlife. Similar comments were made following the Exxon Valdez oil spill (Wiens 1995). At the same time some chemical dispersants may have a longer lasting and more damaging impact on shore-life than leaving the oil to degrade naturally or removing it manually. It also seems that recovery times follow a natural time scale determined by the habitat rather than the method of clean-up, which for rocky coasts is approximately 3 years (Sell et al. 1995).

Figure 4.28. An oiled guillemot

From this it may be deduced that the timing, type and scale of clean-up operations must include an assessment of the real long term threat to the environment against its innate ability to recover. The sight of dead and dying birds covered in oil (Figure 4.28 above) is distressing and the desire to 'do something' may be overwhelming. However, the need to take immediate remedial action during major oil spill incidents on conservation grounds, should be tempered by an assessment of both the efficacy and desirability of treatment.

4.4 Cumulative interactions

Most of the principal conservation issues affecting seabird colonies are relatively well known and common to many areas. A major decline in a particular population is, however, often the result of the cumulative impact of several factors acting together. Examples of such interactions are described next.

4.4.1 Puffin conservation: grazing, oil pollution and sandeels

The puffin feeds exclusively at sea but nests on islands and sea cliffs that are free from predators. It nests in burrows and rock crevices, often occurring in dense breeding colonies. These locations are linked to the availability of food, particularly the large populations of small fish found in the northern waters in summer. Studies in the North Sea show that, as with other seabirds, there are favoured feeding areas. Individuals of this species fish close inshore when feeding their young and congregate only a little further offshore soon after they finish nesting. During mid-winter they become more widely dispersed, extending into the northern seas.

Figure 4.29. Overgrazed puffin habitat on the island of Skomer, Wales. Note the fenced area, free from rabbit grazing and with much more vigorous vegetation than the surrounding area

In recent years the population of the species has declined to a considerable extent at many breeding sites. This appears to be part of a long-term decline attributable to a number of factors acting upon the population when the birds are most vulnerable, during and soon after breeding. When breeding, over-use of burrows may cause them to collapse. This can be particularly important on islands where overgrazing by domestic stock (notably sheep) and/or introduced rabbits may cause soils to become unstable (Figure 4.29 above). In extreme cases the site can be rendered unsuitable for further nesting as the surface soil is washed away.

Infestations of rats have also been implicated in reducing breeding success as young birds are killed and eaten. Oil pollution at sea can also kill large number of birds and may have contributed to the long term decline of colonies along the north coast of France. Taken together these pressures seem to have caused a reduction in the population overall (Harris 1984). More recently, depletion of fish stocks such as sandeels (*Ammodytes*) and sprat (*Sprattus sprattus*), have also been implicated. The fishery around Shetland peaked in 1982 but was closed in June 1990 due to a collapse in the numbers of sandeels and lack of recruitment to the population of puffins. Pressure was also mounted from conservation groups, who stressed that the fishery may have been adversely affecting the breeding success of the seabirds (Bailey 1991, Monaghan 1992). Subsequent research has shown that large fluctuations occurred in sandeel abundance after the closure of the fishery, suggesting that a causal link may not have been involved (Wright & Bailey 1993, Wright 1996). The fishery was re-opened at the start of 1995, but with restrictive management measures.

This example suggests that when considering the conservation of individual species of seabird it is important to recognise that there are often several interrelated factors affecting population numbers. In the case of the puffin its survival at individual breeding sites may depend on decisions taken at local level in relation to the status and management of soils. Where domestic stock graze nesting areas, national or international policies which encourage greater stocking densities (such as the European Union Common Agricultural Policy) may increase the negative impact on the population caused by soil erosion. At the same time over-fishing (regulated through fisheries policies) may reduce breeding success. When these effects coincide with deaths due to oil pollution then the cumulative impact may contribute to the loss, not only of individual birds but of whole populations.

4.4.2 Fishery management

Over-fishing of some species of smaller fish upon which seabirds feed is only one of the effects that fisheries have on seabirds. Fisheries management exerts more subtle and indirect effects on the community structure of some populations. For example up to half the fish caught by fishing vessels is routinely thrown overboard as ‘discards’ of undersized fish and offal. These provide easy pickings for many seabirds and there has been an increase this century in numbers of scavenging species such as fulmars, gannets and gulls in the North Sea (Tucker & Evans 1997). Discards from local fisheries around the Ebro Delta, Spain seem to have been responsible for the increase in Audouin's gull in the area (Oró & Martínez 1994). A reduced fishing effort brought about by changes in Fisheries Policy both in Europe and elsewhere, to improve stocks, would reduce this source of food.

Reduced fishing effort may also have other implications especially when predatory fish species are involved. Successful control of the current over-fishing of species such as swordfish, tuna, adult herring and mackerel could lead to a reduction in the stocks of smaller fish upon which seabirds feed (Tucker & Evans 1997). This might be considered a price worth paying for improvements in the sustainability of fisheries, if balanced by an increase in recruitment and availability of small fish. However rapid changes could lead to the already high numbers of scavenging birds preying on other, more vulnerable seabirds. Taken together, these effects could further deplete some rare seabird populations. Complex interactions such as these need to be taken into account if other serious seabird conservation problems are to be avoided.

Birds are also killed as a direct result of fishing activities. They may be caught in the nets, drift nets being particularly damaging. Long-line fishing with multiple hooks kills species such as the albatross. This is particularly prevalent in Japan (Brothers 1991). Other fisheries impacts include bottom fishing and increased boat traffic. To these should be added toxic pollution, nutrient enrichment, coastal erosion and subsidence which have all been cited as contributing to the cumulative effects on seabird numbers. The discovery of such complex interactions has given us a much better understanding of how fishery activity can affect populations of vulnerable seabird species. In this context, the adoption of the precautionary principle, when dealing with fishery management or changes in agricultural policy, may prove essential to the maintenance of some seabird populations.

4.4.3 Impacts on surrounding habitats

Nesting seabirds do not occur in isolation from their surrounding habitats. Impacts around nesting colonies may involve the removal of vegetation for nest-building and eutrophication. These can have a significant effect, as for example in the case of cormorant (*Phalacrocorax carbo*) which began breeding on the Estonian island of Tondirahu in 1986. The number of pairs rose to a maximum total of 1,690 by 1994. High nutrient levels resulted in the destruction of the vegetation over the whole island such that its silhouette became unrecognisable (Ratas & Nilson 1997). In extreme cases the impact can be significant enough to render the habitat unsuitable for the nesting species themselves.

These effects are also manifested in Europe by the classification used to define Natura 2000 sites (European Commission 1999c). The "Vegetated sea cliffs of the Atlantic and Baltic coasts" includes two communities recognised from Great Britain in the National Vegetation Classification (Rodwell 2000) specifically associated with seabird enrichment, namely:

- *Atriplex prostrata-Beta vulgaris* ssp. *maritima* seabird cliff community;
- *Stellaria media-Rumex acetosa* seabird cliff community.

In Croatia some of the more important diversified rock scrub of the megacliffs depends on guano fertilisation. Elimination of much of the cliff avifauna by hunting and poisoning was followed within a decade by the disappearance of the scrub and its associated animals (Lovric 1993a). From this it appears that seabird populations can have both positive and negative effects on their surroundings. From a management perspective these need to be included in the already complex matrix of issues to be considered when assessing the management needs of these species and their habitats.

5. SALTMARSH

5.1 Introduction and scope

Saltmarshes can be defined by habitats containing halophytic plant communities and associated animals which are tolerant of sea water. Typically they are regularly inundated by the tide, have sometimes rapid accumulations of sediment and, in the absence of enclosure, include transitions to non-tidal vegetation. They are at their most extensive in the northern hemisphere in areas subject to high tidal range (macro-meso tidal) where flat tidal plains develop in the shelter of estuaries and on open coasts protected by offshore sand bars and other structures. A key to their sometimes extensive formation is the presence of an abundant supply of fine sediment which settles out from the water column as tidal movement and wave action are reduced. Pioneer salt tolerant plants such as *Salicornia* spp., *Suaeda* spp. and *Spartina* spp. are among the most frequent colonists in northern latitudes.

Saltmarshes are often described in terms of their vegetation succession (see for example an early account by Chapman 1964). Tidal influences and soil conditions create salinity gradients which impose constraints on individual plants, resulting in a restricted range of species able to tolerate the conditions at least in the early stages of development (Ranwell 1972a). Large saltmarshes may have transitions to other habitats including sand dune, swamp, scrub and woodland. Their high productivity, the range of animals which rely on them for part at least of their life cycle and the complex mosaics of spatial and temporal communities, can result in a high diversity and considerable nature conservation value which belie the relative simplicity of their origins. Packham & Willis (1997 Chapters 4 and 5) provide a description of saltmarsh development and community dynamics.

Though many saltmarshes appear, and are often described in terms of their natural development, they have a long history of human use. In temperate regions grazing and enclosure are the most frequent human activities affecting them. Grazing affects vegetation structure and the type of animal and stock density can have a major influence on plant and animal species diversity. Changes in the historic patterns of use can result in significant alteration to their conservation value which may require direct intervention to restore the former interest. By contrast enclosure for industrial purposes and large-scale arable cultivation when imposed directly on saltmarsh causes permanent loss of habitat (Chapter 11). Where the saltmarsh is 'squeezed' by a fixed embankment, especially in areas where relative sea level is rising, a narrowing of the foreshore may occur and erosion can become a dominant force.

This chapter is concerned principally with the management of saltmarshes of temperate regions of the northern hemisphere. These do not occur in isolation as they depend upon, and form an integral part of, the estuaries in which they are most frequently found. Hence when considering management, the information presented here should be looked at in conjunction with that presented in chapter 10 which deals with wetlands (estuaries, deltas and lagoons) and chapter 11 (grazing marshes and salinas).

5.1.1 Habitat definition

The establishment of halophytic (salt-tolerant) plants on sheltered tidal flats is the first stage in the colonisation of this highly specialised community. Accumulations of sediment around the first colonisers allow other species to invade and the process of succession takes place, leading to sometimes complex stands of vegetation which develop out of reach of all but the highest tides. The details of this process can be found in a number of text books - see for example Ranwell (1972a), Adam (1990) and Packham & Willis (1997). Saltmarshes are, as their name implies, habitats dominated by species which are tolerant of inundation by saline water. A definition is:

"Natural or semi-natural halophytic grassland and dwarf brushwood on the alluvial sediments bordering saline water bodies whose water level fluctuates tidally or non-tidally." (Beeftink 1977)

This definition also encompasses halophytic communities occurring in micro-tidal seas such as the Baltic and Mediterranean, on sandy substrates in transitional zones adjacent to sand dunes and brackish water ditches and dykes in areas of coastal grazing marsh (dealt with in Chapter 11) and in saline lagoons (included in Chapter 10). It is less relevant to saltmarsh communities which occur on rocky shores where sediment deposition is virtually absent or on exposed cliffs where saltspray is driven inland, which are also included here. Inland halophytic communities, associated with salt lakes or upwelling from underground salt deposits and salt steppes, are not covered in this book.

5.1.2 Habitat type

Saltmarshes rarely develop in isolation. Four main types of saltmarsh have been recognised in the literature - see for example Beeftink (1977) and Long

& Mason (1983) - which reflect the linkages with other coastal habitats and the nature of the tidal regime in which they occur (Figure 5.30). They are:

- Estuarine marshes - typical saltmarshes within estuaries where maximum sedimentation takes place to create extensive successional marshes;
- Lagoonal - where a narrow spit of sand or shingle encloses a tidal water body which has a narrow opening to the sea, reducing wave energy and tidal amplitude;
- Beach Plain - exposed or partially protected beaches where sediment deposition is limited and may include wind-blown sand. These can be intermediate between saltmarsh and sand dune;
- Barrier Island - where offshore shingle or sand dune islands enclose and protect tidal land to leeward.

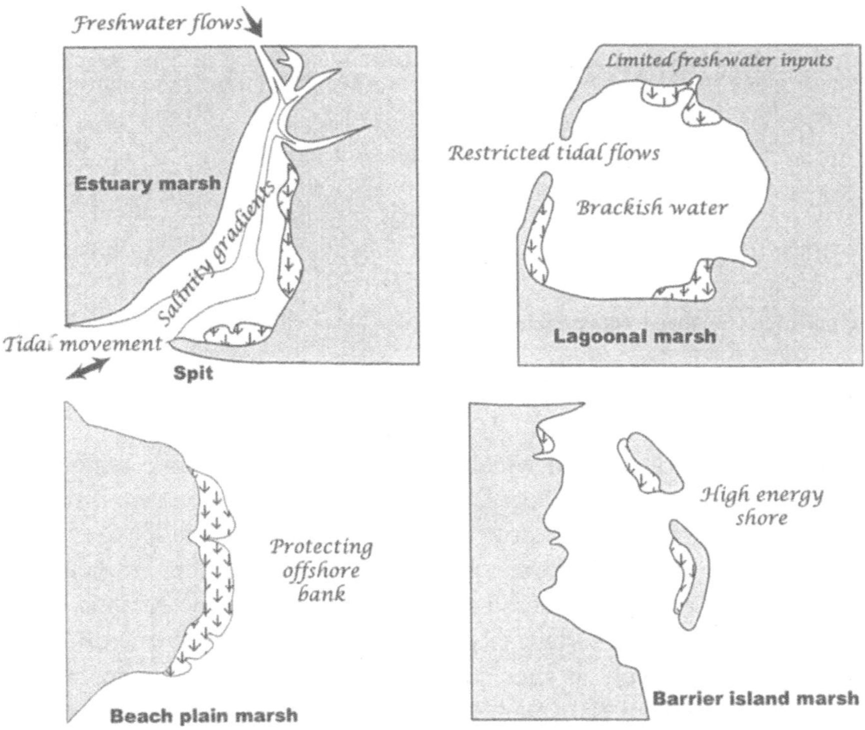

Figure 5.30. Typical sedimentary saltmarshes, based on examples from Great Britain

Abundant sediment availability is a requirement for the development of most saltmarsh communities. However, they also occur in areas where sediment is restricted or absent and can be relatively small in area. A fifth type, '**loch-head marsh**', occupies the margins and heads of sea lochs in the west of Scotland, Scandinavia, northern Canada and Greenland. A definition of these saltmarsh types is also given by Packham & Willis (1997, p. 89). To these can be added '**beach-head saltmarshes**' which exist as a narrow zone on rocky shores where there is little or no successional development. '**Perched**' saltmarsh communities also occur on some exposed cliffs. As waves break against the shore, salt spray is thrown high into the air drenching the vegetation to form communities dominated by salt-tolerant species. They are known from western Ireland, Scotland, Brittany and Portugal (Westhoff 1985), but see also the megacliffs of the northeast Adriatic (Chapter 3). None of the last examples is large: however they are also part of the natural variation in saltmarshes (Figure 5.31).

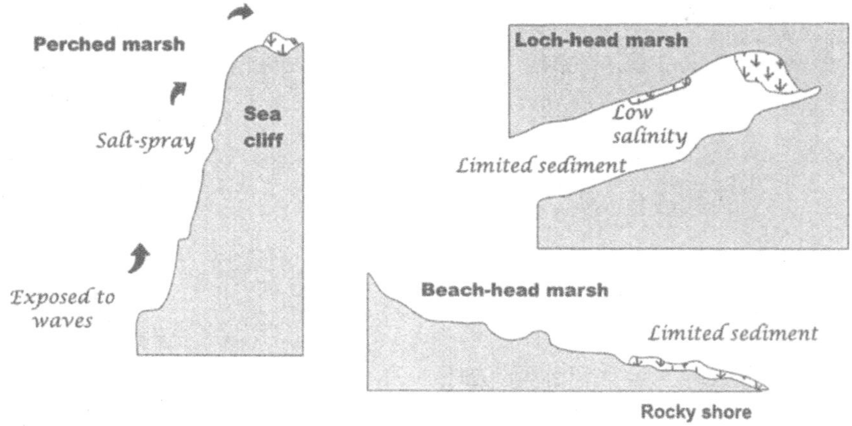

Figure 5.31. Saltmarsh communities, developed in areas with limited or no sediment supply

5.2 Habitat distribution

Saltmarshes are distributed widely throughout the temperate regions of the world (Chapman 1974). In the Arctic and in Boreal regions ice-action and a limited sediment supply restrict their occurrence to small, scattered sites (Adam 1990). They are at their most extensive in the northern hemisphere, particularly in the major macro/meso-tidal estuaries of the Atlantic coasts of northwest Europe and eastern USA. In warmer latitudes in micro-tidal areas they occur extensively at sites such as the Côto Doñana in Spain and the Camargue in France (Dijkema 1984). They are also found in association with the deltas and lagoons of the Mediterranean (e.g. along the northern Albanian coast) and the Black Sea (e.g. in the delta of the Danube). They are replaced by mangrove swamps in the tropics.

5.2.1 Habitat distribution in Europe

In Europe saltmarshes stretch from the Arctic through Scandinavia to the Mediterranean (Figure 5.32). They are at their most extensive in the macro-meso tidal estuaries of Britain and in the Danish, German and Dutch Wadden Sea (Dijkema 1984).

Figure 5.32. Distribution of coastal saltmarshes in Europe (after Dijkema 1984) with additional information. Sites mentioned in the text are named

5.2.2 Habitat distribution in eastern USA

On the eastern United States coastline there are estimated to be approximately 600,000ha stretching from the Arctic where mostly small grassy marshes predominate, through SE Canada, north-east USA, tucked into sheltered coves and bays, to extensive *Spartina* spp. marshes behind the barrier islands and estuaries such as Chesapeake Bay, on the mid-Atlantic coast where there are 100,000ha. Further south in the warmer waters of the

Gulf of Mexico more than 2 million hectares are present (Dardeau et al. 1992); again *Spartina* marshes predominate in tidal areas.

5.3 Nature conservation value

The nature conservation value of saltmarsh is derived from a number of key features. These include the intrinsic importance of the specialised and sometimes rare plants which make up the communities, together with a variety of animals which rely on the habitat for part or all of their life cycle. The plants include a range of salt tolerant species, many of which are restricted to tidally influenced areas. Far from lacking diversity, as is often depicted, in the absence of enclosure there is a wide variety of plants ranging from salt tolerant halophytes to species of upper marsh levels which are predominantly terrestrial. Superimposed on this physical structure of the marsh is the diversity of plant communities whose spatial and temporal variation provides a wealth of specialist niches for many animals. Each of the main interests is dealt with in turn below. The relationship between these interests and the wider estuary environment are considered in more detail in chapter 10. Some of what follows is based on a chapter prepared by the author as part of a guide to saltmarshes and their management in Great Britain (Toft et al. 1995).

5.3.1 Plant community successions and transitions

The apparent simplicity and 'naturalness' of the plant communities has made them the subject of a considerable literature associated with unravelling the basic ecological principles of plant succession (e.g. Chapman 1938, 1939, 1941, Ranwell 1964a, b, 1972a). Several of the early descriptions are derived from studies of marshes in sheltered locations, behind barrier islands where the saltmarsh has developed on a high, often sandy beach plain (Scolt Head, North Norfolk). Successional development does take place and a sequence of vegetation zones running roughly parallel to the shore is frequently found, particularly in the earlier stages of saltmarsh development. However the interpretation of this zonation and the inference that vegetation develops as a steady progression, as the height of the surface is raised by sediment deposition, is a much simplified view of the processes involved (Ranwell 1972a, Adam 1990).

The concept of succession is nevertheless a useful basis upon which to begin to consider the complexities of saltmarshes, their nature conservation importance and management, since it implies a dynamic change in time and space. An example of saltmarsh zonation is shown for an estuarine coast in northwest England (Figure 5.33). Flooding by tidal waters is a natural and essential part of the processes. All species which occur there are to a greater or lesser extent adapted to survive the periodic inundation with sea water, changes in salinity and desiccation when the tide is out.

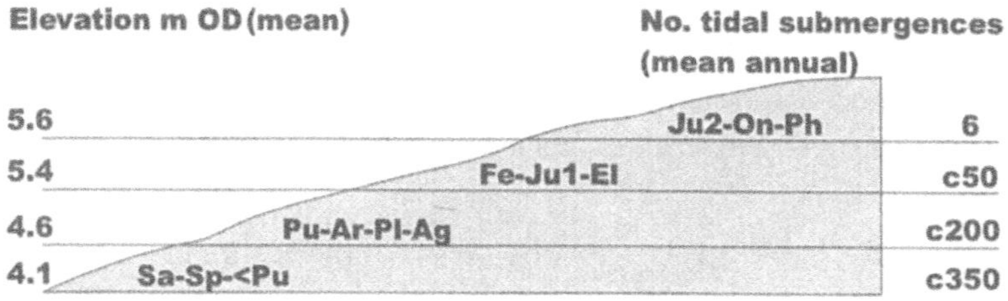

Figure 5.33. Saltmarsh zonation & tidal height. Redrawn from Gray & Scott (1987)

Ph - *Phragmites australis*
On - *Oenanthe lachenalii*
Ju2 - *Juncus maritimus*
El - *Elytrigia atherica*
Ju1 - *Juncus gerardii*
Fe - *Festuca rubra*

Ag - *Agrostis stolonifera*
Pl - *Plantago maritima*
Ar - *Armeria maritima*
Pu - *Puccinellia maritima*
Sp - *Spartina anglica*
Sa - *Salicornia* spp.

It is usually assumed that higher plants play a vital role in the stimulation of accretion on mudflats where saltmarsh develops. The tendency for sediment to settle out is aided by some macro-algae such as fucoids and *Vaucheria* spp., (Carey & Oliver 1918), or micro-algae (Coles 1979). However the precise mechanism is unclear and it may have more to do with preventing re-suspension of sediments than promoting accretion. The initiation of marsh growth thereafter is by pioneer plants which slow the tidal currents, can withstand tidal buffeting and are salt tolerant (Ranwell 1972a).

Descriptions of zonation, as depicted in Figure 5.34, show a straight forward progression from one stage to the next. However, this apparently simple picture hides a sometimes complex relationship. Sequences of erosion may be followed by regrowth as estuary channels change their course. A series of steps can form as new saltmarsh develops to seaward of the eroding cliff. Salt pans can introduce yet another change into the complex mosaic and deposits of seaweed on the tide-line may smother the surface vegetation creating further spatial variation as the strandline deposits rot (Packham & Willis 1997, pp. 101-5). This horizontal mosaic is further diversified by the vertical structure of the vegetation which adds another dimension to the saltmarshes' ability to support a wide range of animals, especially invertebrates (Figure 5.34).

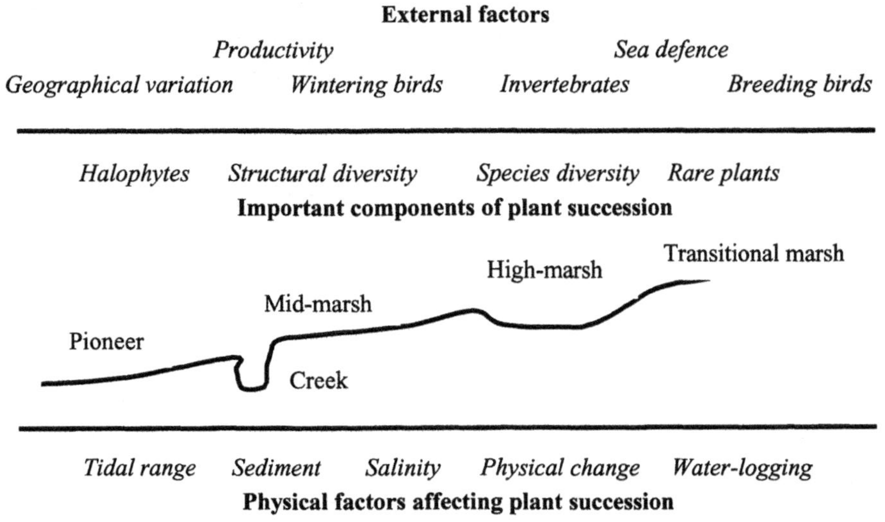

Figure 5.34. Components important to saltmarsh plant and animal diversity

Flows of fresh water may have a deleterious impact on the salt tolerant plants. It can also introduce a further element of diversity on to the marsh and natural freshwater run-off increases the plant community variation. Another variation occurs along the gradient of an estuary. On the eastern coast of the United States of America, the vegetation, which is dominated by two plant species (*Spartina alterniflora* and *S. patens*) in saltmarsh grades into 'brackish bay marsh' and 'fresh bay marsh' on some of the more extensive sites. Complex communities occur across the salinity gradients with a rich variety of plants and animals as in Chesapeake Bay (White 1989).

Saltmarsh plant community mosaics and within estuary successions and zonations are only one part of the importance of saltmarsh vegetation. From a nature conservation point of view the **geographical range** of variation is also significant. Adam (1990) suggests a classification which recognises 10 major types world-wide.

5.3.2 Plant community variation

The world distribution of saltmarsh vegetation shows a relatively simple pattern of variation with a major axis correlated with latitude. Arctic saltmarshes are generally grass-dominated with a low species diversity, small size (due to the lack of fine-grained sediment) and instability. The low marsh is characteristically dominated by *Puccinellia phryganodes*. They are widespread though the effect of ice action and lack of sediment restricts their occurrence. Annuals including *Salicornia europaea* agg. and *Suaeda maritima*, which are common in low levels on temperate marshes, are usually absent.

Baltic saltmarshes span the range of the Boreal marshes and include communities from brackish water to those more typical of the Atlantic. In the north salinities are low and the rocky coasts are subject to isostatic uplift. There is also a limited sediment supply so saltmarshes are small and similar to those of the Arctic. Larger saltmarshes comparable to those in the North Sea are found only in the south especially associated with barrier islands on German and Polish shores (Dijkema 1990). The plant communities lack species such as *Spartina anglica* and *Atriplex portulacoides*, important components of marshes further south and west on the Atlantic coast, which are limited by cold climates.

Saltmarshes in Great Britain vary in composition from north to south. This variation provides a link between the saltmarshes of the Mediterranean (including the 'Dry coast' type of Adam 1990) and the Arctic. In addition there are east/west gradients which help to define the three main types characteristic of the region (Adam 1978, 1981). Type A marshes dominate in southeast England and are characterised by the presence of *Salicornia europaea* in the low marsh and *Atriplex portulacoides* and *Spartina anglica* in the mid-marsh. Type B marshes found in northwest England and Wales, are more grassy and have an abundance of *Juncus gerardii* in the mid-upper levels. Type C marshes of western Scotland have a greater proportion of upper marsh communities but fewer halophytes. The difference between the east and west marshes is attributed to a combination of the heavier grazing and sandier substrate in the west. The presence of *Frankenia laevis* and *Suaeda vera*, in North Norfolk, help define links with saltmarshes of the Mediterranean where these species are widespread.

These relationships form an important part of the evaluation of conservation interests. They are also important to the identification and selection of sites for conservation protection such as the statutorily protected Sites of Special Scientific Interest in Great Britain (Nature Conservancy Council 1989). A more detailed breakdown of types according to the National Vegetation Classification (Rodwell 2000) includes the full range of variation of halophytic, transitional and grassland vegetation found on saltmarshes in Great Britain. At a wider geographical scale the classification of the European Union, Habitats and Species Directive, aims to provide the basis for the selection of habitats of community importance and will lead to national governments identifying a series of sites which will combine to form a European wide network of the most important areas for habitat conservation, 'Natura 2000'. A summary of the main classification of European saltmarshes is given in Table 5.7 (European Commission 1999c).

Table 5.7. Atlantic and Continental saltmarshes and salt meadows. Interpretation Manual of European Union Habitats, European Commission, October 1999. * denotes Priority Habitats

Codes	Directive Name	Geographical distribution
15.1	*Salicornia* and other annuals colonising mud and sand	Belgium, Denmark, France, Germany, Greece, Ireland, Italy, Netherlands, Portugal, Spain, Sweden, United Kingdom.
15.2	*Spartina* swards (*Spartinion maritimae*)	Belgium, Denmark, France, Germany, Greece, Ireland, Italy, Netherlands, Portugal, Spain, United Kingdom.
15.3	Atlantic saltmeadows (*Glauco-Puccinellietalia maritimae*)	Belgium, Denmark, France, Germany, Ireland, Netherlands, Portugal, Spain, Sweden, United Kingdom.
15.4	Inland salt meadows (*Puccinellietalia distantis*)*	France, Germany, United Kingdom. (Not included here)
15.5	Mediterranean salt meadows (*Juncetalia maritimi*)	France, Greece, Italy, Portugal, Spain, United Kingdom.
15.6	Mediterranean and thermo-Atlantic halophilous scrubs (*Sarcocornetea fruticosi*)	France, Greece, Italy, Portugal, Spain, United Kingdom.
15.7	Halo-nitrophilous scrubs (*Pegano-Salsoletea*)	France (Corsica), Italy (Sicily, Sardinia), Spain, Portugal.
15.8	Salt steppes (*Limonietalia*)*	Mediterranean coasts and Iberian peninsula.

The saltmarshes of the east coast of North America are dominated in the lower marsh by *Spartina alterniflora*. Further up the marsh is *Spartina patens*, a lower growing less vigorous plant, which is more prevalent in the north. Despite major enclosure affecting large areas of marsh, extensive transitions to brackish and freshwater marsh occur which are very rich in species of plants and animals. Adam (1990) provides a detailed description of saltmarsh vegetation world wide. The natural distribution patterns are important to unravelling natural changes, such as those brought about by climatic variation, and those caused by human activities.

5.3.3 Rare plants

Saltmarshes are not normally noted for the presence of rare species. Indeed the apparent simplicity of the species communities is often one of the reasons for studying their successional characteristics. However the species list presented for Great Britain (Table 5.8) shows there are a number of rare and scarce plant species found predominantly on saltmarshes. The fact that the majority of these are located in southeast England is a reflection of a general picture of saltmarsh which tend to be more species rich the further south they are situated. The significance of upper saltmarsh levels including transitions to brackish marsh and grazing marsh can also be seen.

Table 5.8. Rare (R) and scarce (S) saltmarsh species in Great Britain

Species	Distribution	10km^2	Status	Habitat
Limonium bellidifolium	eastern	5	R	High level sand/mud
Spartina alterniflora	south	1	R	Low level mud flat
*Atriplex pedunculata**	east	1	R	Grazing marsh
Atriplex longipes	scattered	7	R	Upper marsh
Alopecurus bulbosus	south	26	S	Saline meadows
Suaeda vera	south	30	S	Drift line/shingle ridges
*Bupleurum tenuissimum**	south	58	S	Upper marsh/grazing marsh/sea walls
Inula crithmoides	south	62	S	Upper marsh
Limonium binervosum	south	70	S	Upper marsh/dune transition
Althaea officinalis	south/west	73	S	Upper marsh/ditches in grazing marsh
*Polypogon monspeliensis**	south east	18	S	Enclosed land/grazing marsh
Spartina maritima	south east	25	S	Low marsh
Frankenia laevis	south east	25	S	Upper marsh/sandy flats
Salicornia pusilla	south east	32	S	Low marsh
Salicornia perennis	south east	37	S	Low marsh
Puccinellia fasciculata	south east	52	S	Open saline areas above extreme high water Spring tides
Limonium humile	south	59	S	Upper/mid marsh

*species now confined to grazing marsh in GB (Chapter 11) though formerly present on upper marsh. (The definition of rarity is based on the recorded presence of the species in 10km^2 in Great Britain. Species which occur in 15 or fewer 10km^2 are "nationally rare" and those occurring in 16-100 10km^2 are "nationally scarce".)

5.3.4 Invertebrates

Saltmarshes present difficult environments for colonisation by invertebrates, due to changes in salinity and humidity caused by periodic tidal immersion. The fauna is a mixture of marine, freshwater and terrestrial species adapted in various ways to the stressful environment. Marine species tend to occur lower down the marsh and often burrow to avoid desiccation. These may include few species which occur in great abundance, particularly in the lower marsh. Amongst these the snail, *Hydrobia* spp. and the amphipod, *Corophium* spp. are preyed upon by large numbers of birds feeding in and around many estuarine marshes. Terrestrial and freshwater species occur mostly in the upper marsh and transition zones, and have adapted to, or avoid immersion in saline water. These marsh levels may be rich in invertebrate fauna. Studies of the Wadden Sea saltmarshes by workers from the University of Kiel (reported in Dijkema 1984) suggest, for example, that there may be as many as 1,300 species in the middle saltmarsh.

A sharp discontinuity in faunal distribution occurs at the interface between mud or sand flat and pioneer marsh. This is caused not only by the change in structure, microclimate and food availability, but also by predatory pressure from birds and fish. Species requiring the cover provided by denser vegetation, such as shore crab (*Carcinus maenas*) and rough periwinkle (*Littorina saxatilis*) become more common. Higher up in the mid-upper marsh zones, invertebrates of terrestrial rather than marine origin become more frequent, especially around the better drained margins of creeks.

A further characteristic is the extent to which species are restricted to the different zones of the marsh. This is a reflection of the fact that in the upper and transitional zone species which use the marsh are often those which are equally able to compete in the adjacent terrestrial habitats. Those which occur at lower levels are specialists, restricted to tidally influenced areas.

Groups of species such as shore bugs (Saldidae), ground dwelling predators, include several species which occur on saltmarshes. Each lives in a different part of the marsh. For example *Saldula pilosella* lives at the margins of sheltered pools on the upper shore, *Salda littoralis* in dense vegetation and *Saldula palustris* occurs on more open sandy/muddy areas, extending further down the marsh than the other species (Kirby 1992).

The botanical species richness and structural diversity of an ungrazed or lightly grazed marsh produces a greater spectrum of terrestrial invertebrate niches, ranging from sites for web spinning by spiders, to habitat for specific phytophagous species, such as aphids, weevils, gall midges and Lepidoptera. Whilst most are widespread, occurring in both saltmarshes and adjacent freshwater habitats, a few species that can tolerate immersion in salt water have a competitive advantage. A study of spiders in the UK (Duffey 1970) found two such species *Lasiargus gowerensis* and *Baryphyma duffeyi* in Gower. The litter accumulations at the base of plants are also important habitats for some species. Transition zones to terrestrial habitats support a number of highly specific taxa, notably among the two-winged flies (Diptera). The fauna of brackish water reedbeds is distinct from that of freshwater reed beds, and a discrete suite of species of nocturnal moths, in particular wainscot moths (*Mythimna* spp.) and others, have caterpillars feeding on and in the stems of reed. Reeds, in standing and brackish water, are more favourable for some phytophagous species as they are isolated from ground dwelling predators, which are more prevalent in tidal reedbed or reedbed litter.

The same increase in diversity occurs in the higher and less saline estuarine marshes of the mid Atlantic shores of America (Perry 1985). Ribbed mussel (*Modiolus demissus*) is prominent along tidal banks and marsh snails include oval marsh snail (*Ovatella myosotis*) and the saltmarsh snail (*Melampus bidentatus*) are abundant, providing food for a variety of intertidal predators including many birds. Three species of fiddler crabs inhabit different sections of the marsh: mud fiddler (*Uca pugnax*) and sand fiddler (*U. pugilator*) inhabit tidal marshes and mudflats and more sandy areas respectively, whilst brackish water fiddler (*U. minax*) occurs in fresh or slightly brackish water. Spiders are again important, with web-building species remaining above all but the highest tides. Some wolf spiders which occur on the marsh surface can withstand submergence for several hours. A review of saltmarsh fauna including invertebrates on American saltmarshes may be found in Daiber (1982). Of particular note are the invertebrates such as the numerous species of two-winged fly. Amongst these the biting salt-marsh mosquito (*Aedes sollicitans*), saltmarsh green-head fly (*Tabanus nigrovittatus*) and deerfly (*Chrysops* spp.) are most noticeable as they cause considerable discomfort to humans (Teal & Teal 1969). These and other related species play an important role in the history of saltmarsh as they have often been one of the main factors in promoting the drainage and enclosure of marshes in America (Daiber 1986) and elsewhere in the world.

Strandline accumulations of drift litter, with or without saltmarsh species, support various assemblages of invertebrates where humidity and state of decay of the drift material is important. Accumulations of seaweed at high water mark support a wide diversity of specialist flies. One scarce beetle, *Aphodius plagiatus*, in a genus composed otherwise only of dung beetles, is associated with rotting fungi and accumulations of plant litter in saltmarshes and dunes. So, far from having a restricted diversity of invertebrates, unenclosed saltmarshes are likely to have both a diverse and abundant fauna.

5.3.5 Breeding birds

Successful breeding by a variety of birds occurs on the upper levels of unenclosed saltmarshes following the high spring tides in April and May, slightly later than their inland counterparts. Although early nests may be destroyed by high tides, at least one brood is usually possible. Even where flooding affects nests, as long as the eggs remain *in situ* some species, notably redshank (*Tringa totanus*), continue incubation and the eggs may be successfully hatched. Although other species are not so resilient there are important breeding bird populations on the upper marsh.

The principal species which breed on saltmarshes in northwest Europe are shown in Table 5.9. The structural diversity of the marsh is of crucial importance to some species such as redshank and particularly high populations may be present in the best habitats. Table 5.9 also gives an indication of the main habitat requirements for some of the most common species.

Table 5.9. Breeding birds of saltmarshes and their principal habitats in Great Britain

Species	Habitat
Tringa totanus (redshank)	Tussocky, moderately grazed mid-high marsh
Haematopus ostralegus (oystercatcher)	Moderate to heavily grazed high marsh
Vanellus vanellus (lapwing)	Grazed grassland on high marsh
Larus ridibundus (black-headed gull)	Mounds on marsh; also in *Spartina* islands
Alauda arvensis (skylark)	Upper grassy marsh
Emberiza schoeniclus (reed bunting)	Edge of marsh with tall vegetation
Anthus pratensis (meadow pipit)	Upper grassy marsh

On the eastern shores of America a few bird species are strictly saltmarsh nesting, including clapper rail (*Rallus longirostris*), laughing gulls (*Larus atricilla*) and sea-side sparrows (*Ammodramus maritimus*). These and numerous other species straddle the saltmarsh/brackish marsh interface which provides important nesting habitat. The avifauna is much more diverse than in Europe and this no doubt reflects the extent of the marsh and the survival of full stages of succession and/or transitions to tidal freshwater and brackish water habitats at sites such as Chesapeake Bay.

5.3.6 Wintering birds

Wintering wildfowl including grazing ducks and geese such as wigeon (*Anas penelope*) and brent goose (*Branta bernicla bernicla*) may feed directly on grassy saltmarshes. Other species such as the waders dunlin (*Calidris alpina*) and knot (*Calidris canutus*) feed on the adjacent mudflats in estuaries using the saltmarsh for roosting at high tide or as supplementary feeding around the pans and bare areas on the marsh. Only redshank and snipe (*Gallinago gallinago*) of the wading birds use saltmarsh in preference to the exposed tidal flat, possibly because they like to be sheltered from view by the cover of vegetation. In contrast, geese seek out open areas with close cropped vegetation as they prefer to be able to see potential predators.

Passerines such as skylark, twite, reed bunting, rock pipit and lapland bunting (*Calcarius lapponicus*) are also winter visitors to saltmarshes in Europe and large numbers of individuals can occur. See for example the extensive 5,000ha saltmarshes of the Wash, one of the few estuarine marshes where information is available (Table 5.10).

Table 5.10. Estimates of the most numerous passerines on the Wash (after Davies 1987)

Species:	Estimated Nos. (Winter)
Alauda arvensis (skylark)	26,000 - 32,000
Acanthis flavirostris (twite)	up to 17,000
Emberiza schoeniclus (reed bunting)	500 - 1,000
Anthus spinoletta (rock pipit)	1,200 - 2,300

5.3.7 Other animals

Apart from the occasional fox and other typically terrestrial mammals, saltmarshes are not known for their importance for particular groups of other vertebrates. The rare natterjack toad (*Bufo calamita*) frequents upper levels in northwest England and southern Scotland. In America at least one species, muskrat (*Ondatra zibethicus*), relies on the tidal marshes as a significant source of food and shelter. In the UK otter (*Lutra lutra*) may be attracted to the larger sites because of the relative lack of disturbance. In addition, as they represent some of the largest expanses of uncultivated land smaller mammals such as voles (*Microtus* spp.) and shrews may be abundant on the upper marsh and provide an important source of food for species of bird including short-eared owl (*Asio flammeus*) and kestrel (*Falco tinnunculus*).

Marsh edges and saltmarsh creeks are also important and in some sites the common seal (*Phoca vitulina*) uses these regularly. Breeding colonies occur at many sites in the southern North Sea with especially large concentrations in the Wash, England and the Wadden Sea. They also occur on a few west coast estuaries where they may pup on tidal sand banks or saltmarsh creeks. The interaction between saltmarshes and estuaries is described in more detail in chapter 10. In areas where adjacent land is intensively cultivated or industrialised these saltmarshes may provide the only undisturbed land for some of the commonest creatures.

5.4 Human activities and conservation

Historically many of the human activities which affect saltmarshes today are probably derived from techniques borrowed from natural processes such as accretion, development of creek banks, formation of pans and drainage creeks (Beeftink 1977). By taking advantage of any natural extension of the coastline it was possible to create more favourable conditions for exploitation This firstly involved use by grazing stock and secondly enclosure and subsequent use for agriculture. As engineering techniques improved the loss of saltmarshes increased dramatically.

Throughout Europe and north America saltmarshes have been the subject of major and permanent enclosure for agriculture and together with infilling areas have been developed for ports, industry and housing. Other activities, such as use by domestic stock, whilst less damaging, cause changes to the nature of the marsh and its vegetation, such that it may be difficult to find any saltmarsh that has not been altered in some way by human use. Thus any consideration of management options must take account of this. A list of human activities impinging on saltmarshes are listed below (Table 5.11) and illustrated in Figure 5.35.

Table 5.11. Key management issues taken mainly from examples in northwest Europe

ENCLOSURE, including:	**ACCESS FOR SPORT AND RECREATION**:
Saltmarsh loss (primary land claim);	Bird-watching;
The creation of grazing marsh (Chapter 11);	Walking;
Intensive agriculture (secondary land claim).	Wildfowling;
SEA-LEVEL CHANGE:	Boating/mooring;
Effects on vegetation;	Leisure fishing;
Breeding birds;	Power boating and jet skiing;
The saltmarsh 'squeeze'.	Sailing;
GRAZING MANAGEMENT:	Wind surfing;
Ungrazed;	Horse riding.
Low-moderate grazing;	**REMEDIAL ENGINEERING**:
High levels;	Excavation (of upper saltmarsh);
Formerly grazed marsh.	Impact on rare plants;
OTHER USES:	Protecting the eroding edge;
Turf-cutting;	Polders.
Hay-making;	**POLLUTION**:
Reed thatching;	Oil and Chemical
Samphire gathering	Litter
Spartina planting / control	

The impacts of climate change, and in particular sea level rise, provide a further major series of problems, where landward migration is prevented and erosion takes place at the seaward edge of the marsh. Grazing can also have a profound effect on saltmarsh vegetation and the balance between the two extremes of high and low stocking density is a key management issue.

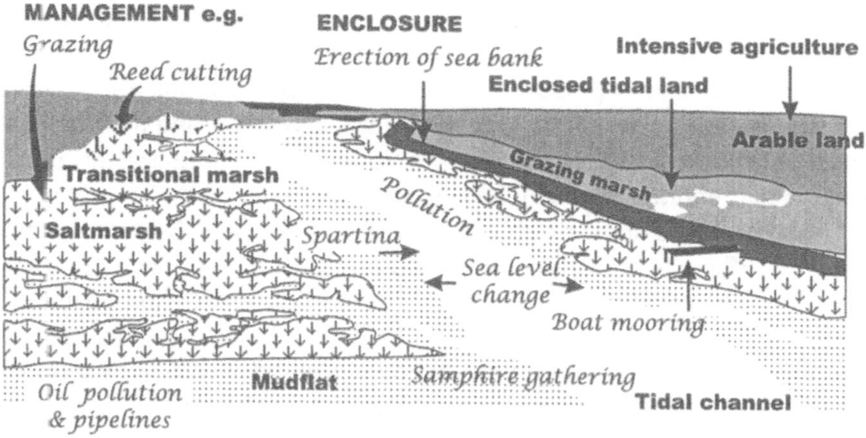

Figure 5.35. Principal causes of saltmarsh loss and important management issues

5.5 Historical losses and sea level change

The enclosure of saltmarsh and transitional swampland represents a subset of the changes which have taken place throughout the world in coastal wetlands (dealt with in Chapter 10). Even in the absence of permanent enclosure for agriculture, industry or other urbanisation, drainage of upper zones, including swamps, was promoted in order to destroy breeding sites for mosquitoes and midges. In eastern USA the saltmarsh mosquito was a particular nuisance and from the 1930s onwards saltmarshes from New England to Maryland were ditched and drained (Teal & Teal 1969). In the Mediterranean around 200 years BC malaria was a major disincentive to settlement in the coastal region south of Rome. From 160 BC a widespread reclamation scheme and re-population was attempted (Torresani 1989). Drainage continued into the 1930s under the direction of Mussolini. Direct habitat loss usually involves enclosure for agricultural use, though in highly industrialised countries the development of ports or other industrial urbanisation may affect both saltmarshes and mud flats.

5.5.1 Enclosure for agriculture (primary land claim)

Enclosure of saltmarsh and development for intensive agricultural use has had a major impact on the habitat. The process involves the exclusion of the tide by the erection of an earth bank or other sea wall, enclosing mature saltmarsh. Early attempts by the Romans took place in the Wash and Severn estuaries. During the 12th and 13th centuries attempts were made to exclude the tide from the upper levels of the saltmarsh in North Norfolk and early enclosure and subsequent use for arable land was commonplace in Kent in the 13th and 15th centuries. These attempts at embankment and drainage probably worked with the natural processes rather than against them and people adapted natural processes to their own ends (Beeftink 1975). In this way they may have had little long term impact on the natural value of the marsh. In historical times these areas were left as rough grassland 'grazing marsh' under traditional grazing management (Chapter 11).

The sequence is similar throughout northwest Europe. Examples include the Lower Elbe where conversion of natural vegetation has been in progress for 2,000 years resulting in loss of most of the tidal areas and transitions to terrestrial vegetation (Garniel & Mierwald 1996). In northeast Germany, the 'Bodden Coast' had extensive areas of grassy saltmarsh shaped by trampling and grazing by cattle over many years. Between 1960 and 1989 large areas were dyked and drained and turned into grassland for intensive cultivation (Lampe 1996).

In the UK elsewhere the change did not take place with one major enclosure but as a series of steps. In the Wash, eastern England, 29,000ha including 3,000ha in 20th century were enclosed over a similar period (Dalby 1957). In the Ribble Estuary, 2,000ha were enclosed in the last century (Royal Commission on Coastal Erosion 1911) with further areas this century (Figure 1.3). In areas where sediment accretion is favoured and in the absence of a rapid rise in sea level, the usual trend is for saltmarshes to extend seawards. In places like the Wash this process (together with *Spartina* planting, Chapter 6) helped continue a tradition of 'winning' land from the sea, which had been carried out for centuries.

Elsewhere in Europe enclosure of saltmarsh appears to be no less dramatic. Although the literature usually quotes losses to wetlands without identifying saltmarsh specifically, it is clear that as component habitats they have suffered as much as those in Great Britain. Musi et al. (1992), for example, describe the losses of coastal habitats within the lagoonal system of Friuli-Venezia Giulia, northeast Italy. Amongst these the saltmarshes associated with the lagoon "have been reclaimed for agriculture or transformed into fish farms". It is impossible to estimate the total area of saltmarsh lost in this way, though it has been suggested that in Europe between a quarter and a half of all saltmarsh may have been destroyed (Dijkema 1984). In the USA Teal & Teal (1969) compare figures in 1922 (based on a survey by the Department of Agriculture) and 1955 (from a survey by the Fish & Wildlife Service) and suggest that a minimum of 2 million acres (at least one quarter) of the shallow coastal water and marsh were lost in the intervening period. More recent figures put this loss as high as 50% (Hinrichsen 1998)

5.5.2 Saltmarshes and sea level rise

Saltmarsh accretion has been estimated to average between 2 and 10mm per annum (Ranwell 1972a) for temperate European and northeast American marshes. These values exceed the current global mean sea level rise of 1-2mm per annum estimated by Gornitz (1993). It is therefore reasonable to assume that new saltmarsh will continue to develop where there is sufficient sediment, and that it can keep pace with sea level rise. However, the rate at which sea level is changing relative to the land is a product of many factors, not least the balance between isostatic and eustatic change, so the response of an individual marsh is difficult to predict. In the case of the Wash, for example, despite the fact that the land is sinking by 1mm per annum, reinforcing the rise in mean sea level (Shennan 1989), saltmarsh continues to accrete (Hill 1988). Up until the 1980s when enclosure of saltmarsh on the Wash effectively ceased, the erection of each new embankment was followed by the growth of new marsh to seaward until a new equilibrium was formed between the position of the sea wall and the edge of marsh (Kestner 1975).

Observations in the saltmarshes in Essex and Kent, only a short distance along the coast suggest this may not always be the normal progression. Here land levels are sinking faster than in the Wash and there has been extensive enclosure over many years. Although large areas of saltmarsh still survive, there is evidence that significant erosion has taken place in recent years (Burd 1992, Table 5.12). It is thought that in the long term the mudflats may become flatter at the expense of the saltmarsh (Pethick 1992).

A similar lowering of beach levels is described for shores fronting sand dunes and sea cliffs during periods when sea level is rising (Bird 1993). In the case of saltmarshes, where the upper limit is constrained by a sea wall the intertidal habitat is squeezed between it and the rising sea. If the rate of sea-level rise increases as a result of global warming, then the saltmarsh zone could ultimately disappear altogether, as it has done already in some localities in Essex where the sea wall is now being undermined (Figure 5.36).

Table 5.12. Loss of saltmarsh in the estuaries of Essex & Kent 1973-1988 (from Burd 1992)

	Original area (ha)	Total area lost (ha)	Loss (ha) to reclamation	Loss (ha) to erosion	% original area eroded
Orwell	99.5	39.9	7.4	32.5	32.6
Stour	264.2	129.5	13.3	116.2	44.0
Hamford Water	876.1	170.6	1.2	169.4	19.3
Colne	791.5	97.7	5.2	92.5	11.7
Blackwater	880.2	200.2	-	200.2	22.7
Dengie	473.8	46.7	-	46.7	14.1
Crouch	467.1	146.1	22.1	124.0	26.5
Thames (Essex)	365.9	105.6	22.3	83.3	22.8
Thames (Kent)	77.8	17.5	3.2	14.3	18.4
Medway	843.8	198.3	18.2	180.1	21.3
Swale	377.0	61.6	3.4	58.2	15.4

Figure 5.1. A seawall being undermined following saltmarsh erosion, Essex

5.5.3 The saltmarsh 'squeeze'

In areas where sea level is rising relative to the land the enclosure of saltmarsh, whether for agriculture or infrastructure development, results in the saltmarsh being 'squeezed' between the development on the land on the one hand and the rising sea level on the other. One obvious effect is the loss of the upper marsh levels and transitions to terrestrial vegetation and with them some of the richest communities. Although cause and effect have not been established, further erosion of saltmarshes can be expected. At the same time the area of saltmarsh is reduced so is its ability to accommodate changes in the external environment, notably through storms and changes to sediment regimes and tidal patterns. Thus conservation management must take into account the extent to which the functioning of the marsh is already compromised by historical losses. In addition, other factors such as saltmarsh productivity or its ability to act as a pollutant sink may be impaired. These may have consequences which affect the functioning of the whole system and ultimately the sustainability of human use.

These results suggest that the assumption that areas of saltmarsh will continue to respond positively to a rising sea level cannot be sustained. The relative change in sea level is an important component in assessing the likely future management requirements on an individual marsh.

5.6 Grazing on saltmarsh

In Europe saltmarshes were almost certainly amongst the first areas to be used for grazing domestic stock. The naturally open nature of the tidally washed turf, probably already grazed by wild animals, could have been particularly attractive in an otherwise densely wooded landscape. Today grazing occurs extensively on marshes in Europe (Dijkema 1984), eastern Canada (Roberts & Robertson 1986) and Japan (Ishizuka 1975). The type of animal includes sheep and cattle in northwest Europe, together with goats in the Mediterranean. Other areas such as the Camargue in southern France also include horses, bulls and the introduced South American *Myocastor coypus* (coypu) (Weber & Hoffman 1970). Although much reduced, in many areas grazing continues and is a major determinant of the structure and nature conservation importance of the saltmarsh habitat.

The levels of grazing are defined by the level of standing crop namely:

- light grazing - most of the standing crop is not removed;
- moderate grazing - standing crop almost completely removed;
- heavy grazing - height <10cm, all standing crop removed;
- abandoned grazing - matted vegetation, no standing crop removed.

These definitions and the grazing levels associated with them which are quoted below are for artificial saltmarshes in the Wadden Sea (Dijkema & Wolff 1983). As such, they lie at the upper range of the levels considered to be appropriate for nature conservation management on unenclosed saltmarshes.

5.6.1 Ungrazed / light grazing

Traditionally ungrazed or lightly grazed marshes are those where native herbivores (grazing ducks and geese, hares and rabbits) are the only grazing animals or where levels of stocking by domestic animals are 2 - 3 sheep or 0.7 - 1.0 young cattle per ha, or lower, for 6 months of the year (normally between the months of April and October). NB Beeftink (1977) recommends 2 sheep or 0.3 cattle per ha (year round) as being the most appropriate to attain the maximum nature conservation interest. These are probably closest to 'natural' marshes which have a complete sequence of vegetation from pioneer to strandline and transitions to terrestrial habitats. In addition they tend to have a good structural diversity and support plant communities with a number of grazing sensitive species such as *Atriplex portulacoides*, *Limonium vulgare* and *Artemisia maritima* (Figure 5.37). These species are each important constituents of the vegetation and support a wide range of invertebrate animals which feed or find shelter on the plants themselves. Breeding birds also find the greater structural diversity provides shelter for nests.

Figure 5.37. Ungrazed saltmarsh, North Walney, Lancashire, England

General recommendation:
Traditionally ungrazed or lightly grazed saltmarshes - continue existing regime. Do not introduce grazing on marshes ungrazed by domestic stock, though see below "abandoned (formerly grazed)".

5.6.2 Moderate grazing

Ungrazed marshes are the exception in northwest Europe. Most marshes have been used for grazing domestic stock at least at moderate levels. These are usually given as being equivalent to 5 - 6 sheep or 1.0 - 1.5 young cattle per ha (April - October). These figures are somewhat higher than the recommended levels given by Beeftink (1977) and suggest that at these levels nature conservation value may be reduced.

Moderately grazed saltmarshes will have a variety of nature conservation interests depending on the type of grazing animal (for example cattle will produce a more structurally varied vegetation mosaic than sheep). Timing of grazing (summer only, winter only, or all year), density of animals, presence of native herbivores (grazing ducks and geese), availability of adjacent non-tidal land and other uses such as turf-cutting will all influence the status of conservation interest. Higher stock levels will favour wintering waterfowl at the expense of some of the other conservation interests. In Britain many phytophagous invertebrate species are associated with plants that are sensitive to grazing. Lower stocking rates, especially if cattle are used, will favour greater structural diversity and encourage interests including breeding birds. Although cattle may damage nests and disturb breeding birds at the lower stocking levels this may not be a severe problem. How these affect conservation interests will depend on the time of year when the grazing takes place.

General recommendation:
Saltmarshes with current stocking levels ranging from low intensity regimes to moderate stocking rates provide a range of opportunities for conservation management (see below).

5.6.3 Heavy grazing

Saltmarshes which have a long history of heavy grazing by domestic stock lack certain attributes such as a rich and varied flora and fauna associated with marshes with greater structural diversity. Grazing levels equivalent to 9 - 10 sheep or 2 - 2.5 young cattle per ha (April - October) given by Dijkema & Wolff (1983) approach those of inland grassland. In northwest England where some of the most extensive and intensively grazed sites occur, stocking densities up to 6.5 sheep (year round) or 2 cows (summer) per ha.

occur (Gray 1972). At these levels, except at the very upper limits of unenclosed marsh such as in Morecambe Bay and in the Solway Firth where transitional upper saltmarsh - mesotrophic grassland may be relatively rich in species, there is an impoverished flora over much of the rest of the marsh.

Grazing sensitive species such as *Limonium vulgare* and *Atriplex portulacoides* are eliminated and tillering grasses favoured. At the same time the structural diversity of the marsh is reduced as all of the standing crop is removed. Over time a close-cropped sward is produced (Figure 5.38) with a much reduced invertebrate fauna. The lack of animal diversity associated with more moderately grazed sites is countered by the presence of often large numbers of wintering ducks and geese. A few species, such as the oystercatcher, breeding in more open locations can also be found associated with these heavily grazed sites.

Figure 5.38. Heavily sheep-grazed saltmarsh, western Ireland

General recommendation:

Traditionally heavily grazed marshes, particularly where high sheep stocking rates are employed, favour short swards and benefit winter grazing ducks and geese. A reduction in grazing levels will increase the potential value of the marsh for other interests but these must be set against the loss of suitable grazing for wildfowl.

Typical values taken from saltmarshes grazed by domestic stock on nature reserves in Great Britain give an indication of the most appropriate regimes for reserves managed for wintering wildfowl (Table 5.13). It is notoriously difficult to determine the appropriate stocking rates for individual marshes. The figures can provide only a guide as grazing may also take place on adjacent grazing land behind the sea wall, to which cattle may also have access. These areas are generally not included in estimates of available grazing land.

Table 5.13. Sites for which grazing levels have been set to support grazing ducks (wigeon*) and geese (barnacle goose**) on nature reserves in Great Britain

Site	Area marsh (ha)	Stock	Number	Period	Density per ha.
Ribble Estuary*	809	Cattle	1,100	May-Sept	1.5
Mersey Estuary*	350	Cattle	500	May-Sept	1.5
		Sheep	650	Sept-Dec	1.8
Dee Estuary	600	Sheep	2000	Summer	3.3
Bridgwater Bay*	36	Sheep	80-150	April-Oct	2.2 - 5.1
Caerlaverock**	97	Cattle	195	May-Sept	2.1
Rockcliffe	875	Cattle	800	May-Sept	1.1
Swale	80	Sheep	250	All year	3.0

5.6.4 Abandoned (formerly grazed)

Grazing of saltmarsh by domestic stock has been abandoned on a number of formerly heavily grazed sites. This is generally the case over the last 30 years in Holland (Dijkema 1984) and around the Baltic (Dijkema 1990, Lundberg 1996). The effect of this is to allow the species favoured by grazing, which are often some of the most vigorous and held in check by the grazing pressure (Gray & Scott 1987), to become dominant. This abandonment is characterised by litter accumulation, dominance of a single species and reduced species diversity and matted, coarse-grained vegetation mosaic (Bakker 1985). After only a few years plant species diversity is further reduced and the saltmarsh becomes less suitable for invertebrates, breeding and wintering birds.

General recommendation:

Where grazing has been recently abandoned it should be reintroduced. Whilst high initial stocking levels may be needed to open up the dense turf, the most diverse conservation interest will be attained by using low level open range grazing regimes identified above.

5.6.5 Deciding on a grazing regime

The current status of the marsh vegetation and animal interest is the first stop in assessing future conservation management. These will usually have been determined by the existing interest including the presence of important plant communities, rare plants and animals, breeding birds and/or wintering ducks and geese. Before changing a grazing regime it is crucially important to have some knowledge of the historical pattern and its relationship with other uses such as turf cutting. As a general principle, unless there has been a major change in management in recent years, maintenance of the status quo should be a first option until investigations reveal the nature of any adverse changes or opportunities for improving existing management.

In summary, low stocking levels may be most similar to natural grazing regimes, providing a structurally diverse vegetation with high breeding densities of some birds and a rich invertebrate fauna. As stock densities (of sheep and cattle) increase there is a shift in nature conservation value, towards the creation of low-growing open grass swards lacking structural diversity. In the early and middle stages of succession the plant communities tend to be impoverished as grazing sensitive species are eliminated. The loss of structural diversity also adversely affects the invertebrate and breeding bird populations. By way of compensation, these areas can provide palatable herbage for large numbers of grazing ducks and geese which preferentially graze the smaller grasses.

By increasing or decreasing stock densities the vegetation can be pushed to favour one interest over the other. In considering what action to take it is important to recognise that a change in conservation status can take place very rapidly. A marsh seemingly supporting a rich and varied flora and fauna can show a loss of interest in a matter of only a few years. A *Puccinellia maritima* dominated sward, for example, which had persisted for at least 50 years reverted to a dense *Elytrigia atherica/Festuca rubra* sward in only 10 years following cessation of grazing (Ranwell 1964b). Grazing ducks and geese can themselves influence the species composition of a marsh, see for example Packham & Willis (1997) p. 39 for an evolving sub-arctic saltmarsh. Wigeon preferentially grazed *Puccinellia maritima* and *Agrostis stolonifera* to the benefit of the less palatable *Festuca rubra.* Such situations can be reversed both by mowing and the reintroduction of grazing stock (Cadwalladr & Morley 1971 and Cadwalladr et al. 1972). However, although mowing increases species diversity initially, a turf of *Festuca rubra* is produced and a decrease in plant diversity occurs after only 5 years. In contrast, re-introduction of cattle grazing alone enhanced species diversity and was attributed to gradual litter decomposition by trampling. On the east coast of the United States *Anser caerulescens* (snow goose) may locally put enough pressure on a saltmarsh to remove most of the vegetation over just one season (Daiber 1982).

From this it is clear that any change in grazing should be made in the light of good background knowledge of the current conservation interest and past management. A review of saltmarshes in Europe (Dijkema 1984) provides some indication of grazing levels appropriate to different situations taken from saltmarshes in the Wadden Sea (Table 5.14).

Table 5.14. Annual average rates, number per hectare, estuary saltmarshes in the Wadden Sea

Level	Sheep	Young cattle	Dairy cattle
Light	2.0 - 3.0	0.7 - 1.0	0.3 - 0.5
Moderate	5.0 - 6.0	1.0 - 1.5	0.5 - 0.75
Heavy	9.0 - 10.0	2.0 - 2.5	1.0 - 1.25
Upper marsh (manured)	-	-	2.0

These figures (and table 5.13 above) are indicative only. Prevailing weather conditions, previous stocking rates and other environmental considerations must be taken into account. Do animals, for example, have access to the marsh all year round and at all stages of the tide? Is there a link between grazing on the saltmarsh and adjacent farmland, such that the marsh is used only infrequently?

5.7 Other management issues

5.7.1 Turf-cutting

Turf-cutting is known only from a small number of sites in northwest England. The process involves the preparation of the turf by the use of reseeding and fertiliser treatment. This helps to perpetuate the dominance of the short-leaved grasses *Puccinellia maritima* and *Festuca rubra*, which are essential components of the turf. The combination of management for cutting turf and very high sheep numbers represents the most intensive form of management encountered on saltmarshes. Its introduction to saltmarshes where it is not already taking place, is not recommended.

5.7.2 Hay-making

In much of northwest Europe hay-making was a common use (Dijkema 1984). It declined rapidly before the 1960s in areas such as the Baltic (Dijkema 1990). In the eastern USA the early settlers relied on the saltmarsh hay crop which was cut with specially adapted mowers drawn by horses. It remained the most important product of the marsh for some time (Teal & Teal 1969, Daiber 1986). Hay-making has now practically disappeared from Europe where it probably helped to retain open saltmarsh communities. There are a few isolated examples where the practice continues, such as on the Portuguese side of the Spanish border (Reserva Natural do Sapal de

Castro Marim, in the estuary of the Rio Guadiana). Here cattle grazing still takes place and cutting hay by hand was observed in 1997.

5.7.3 Samphire gathering

It is likely that the gathering of *Salicornia* spp., a succulent and tasty plant, which has been practised for centuries, is probably relatively benign. In Great Britain it was recorded from only 9 of the 155 estuaries covered by the "Estuaries Review" (Davidson 1991). There are no known long-lasting impacts either on the vegetation or other nature conservation interests. By its nature collection of the material can take place only at low tide and in the late summer before the main wintering bird populations appear. Birds that are present have a wide intertidal zone to use if disturbed.

5.8 Recreation and access

Until recently the main value attached to saltmarsh has been for grazing, bird watching, research or providing quarry for wildfowlers. Although the 'wilderness' quality of the larger areas is increasingly appreciated (Figure 5.39), saltmarshes are not considered as having high recreational potential. For many people they are seen as dangerous and inhospitable places, best appreciated from the adjacent land. Pressures are therefore not great, and generally speaking access to saltmarshes does not pose a direct threat to nature conservation interests.

5.8.1 Recreation

Most recreational activities listed in Table 5.11 which require access to, or through the marsh have only a localised impact on the habitat itself. Paths across the marsh used continuously may cause compaction of the substrate and can impede drainage. The use of vehicles is more serious, causing the break up of the surface vegetation and changes to drainage which may ultimately result in erosion. At a few sites continuous use of the upper margins of the marsh edge can do proportionally greater damage to the nature conservation interest as it tends to affect the richer transition zone. Changing access points may help solve such problems. Power boats and other uses which create waves may also exacerbate erosion at the saltmarsh edge.

Figure 5.2. Saltmarshes can be wild and beautiful places

Although tides and terrain make direct access onto many marshes both difficult and unattractive, walking at the edge of the open estuary landscape is a common pastime. Normally this has little impact on the saltmarsh itself or other conservation interests. However, birds feeding in the intertidal zone need secure roosting areas at high tide. These may be particularly important to the survival of individual species during periods of bad weather. Disturbance of these high tide roosts could therefore have an impact on some waterfowl if they are continuously disturbed. The energy expended in moving from a roost may significantly affect the bird's ability to survive.

5.8.2 Wildfowling

Saltmarshes have been used for wildfowling for centuries. There is some evidence that in the first half of the 20th Century numbers of waterfowl in Great Britain declined due to hunting pressure (Tubbs 1996). Despite this, hunters have often been active in habitat protection and coastal nature reserves may have shooting zones within them. The main impact on the saltmarsh comes directly from the desire by the hunter to improve the availability of habitat for some quarry species such as grazing ducks and geese. This has led in some areas to the mowing saltmarsh vegetation or digging scrapes to increase the area of pools on the marsh. Both activities may have a deleterious impact on the more diverse upper saltmarshes.

5.9 Pollution

Saltmarshes are often exposed to a variety of pollutants. These may be derived from the run-off from agricultural use, contamination from industry (heavy metals), sewage disposal, oil and litter which is carried into the system via the riverine input. The sea may import litter and more importantly oil from accidental or deliberate discharge at sea. All these forms of pollution are likely to have some impact on saltmarsh. In general, saltmarshes are thought to be significant sinks for pollutants. These can be incorporated via the sediments which accumulate in the saltmarsh, which themselves may be contaminated, and by the direct uptake by the saltmarsh plants.

5.9.1 Oil

Oil is perhaps the most conspicuous contaminant affecting saltmarshes. This may take the form of chronic pollution from discharges by estuary-based oil industries, or by occasional input from oil incidents occurring at sea. A review of major accidental oil spillages during the 1960s and 70s, including the 'Torrey Canyon', suggests that saltmarsh vegetation often recovers well after a single oil spillage (Baker 1979). Chronic pollution on the other hand, may cause long term loss of saltmarsh vegetation. Effects on individual marshes can be very different from one incident to another. This depends on a number of factors which include the fact that some plants are more tolerant to oil than others. In addition the volatility of the oil itself is important, with the lighter more penetrating fractions being the most toxic (Baker et al. 1994).

The presence of oil deposits on a saltmarsh may suggest that its removal should be undertaken as a matter of urgency. Where the crude is heavy and easily picked up it may be possible to remove it mechanically. However, this may cause physical damage to the marsh and is not recommended, except for the heaviest contamination near to, or on the high water line. Evidence from studies in the eastern USA (Lane et al. 1987) suggests that use of chemical dispersants may also be more damaging than leaving the oil to degrade naturally. Other studies lead to the view that the time for ecological recovery of oiled saltmarsh is approximately 5 years irrespective of the type of clean-up operation (Sell et al. 1995).

Studies of oil pollution by Ross et al. (1973) indicated that with arctic plant communities 20-55 % recovery on treated plots took place within 1 year. These workers concluded that there was considerable merit in leaving the contaminated areas undisturbed. Crude oil damage was greater in the summer because oil penetrated deeper into the soil. In general, contamination should be prevented in the first instance. In the event of an oil pollution incident occurring the oil should normally be left to degrade naturally.

5.9.2 Litter

Litter can accumulate on the upper saltmarsh and along the tideline. There are three main sources; the most important is the detached material from the plants of the higher marsh. Algae and smaller plant material, such as diatoms and other detritus, are the other sources. Natural litter, in the form of decaying algae or other vegetable matter, may be important in providing food for shoreline birds. For example the snow bunting (*Plectrophenax nivalis*) feeds on the seeds deposited among the litter on the tideline on the east coast.

The build up of plastic and other man-made, non-degradable material is unsightly and can smother vegetation. This may lead to the impression that the saltmarsh and its adjacent habitats are degraded and unattractive areas suitable only as wasteland which should be 'improved' by land claim. It is significant that the results of the investigation of human impacts on British estuaries (Davidson 1991) showed that no fewer than 122 of the 155 estuaries in the review suffered some form of informal (possibly illegal) domestic refuse disposal and 58 had rubbish tips upon them. There is no information on the long term impact on saltmarsh vegetation of small deposits of litter on the marsh.

5.10 Complex relationships

The apparent simplicity of saltmarshes belies their often complex structure as indicated above. Another important lesson for the conservation manager lies in understanding the needs of some important associated species. Two examples the bittern (*Botaurus stellaris*) and its feeding requirements and *Bufo calamita* (natterjack toad), illustrate some of these complexities.

5.10.1 Reeds, bitterns and eels

The natural pattern of *Phragmites australis* (common reed) colonisation is as a successional stage from brackish-fresh water at the top of the marsh to coastal scrub and woodland. As the reed develops at the edge of the marsh there is a build up of dry litter and the reedbed can eventually be raised above the highest tides. Where specialist species occur in association with reed beds (e.g. bittern) management may be required to arrest the succession. At Titchwell Marsh, North Norfolk old reedbeds were drying out and a 1m high bank was built to retain water and reverse the process. However, unforeseen by the managers, the bank blocked the path of young eels, one of the principal food items of the bittern, a rare and declining species in northwest Europe. One consequence appears to have been a further decline in the breeding bittern population at the site during 1980 and 1991. This has been correlated with an increase in the average weight of the eels crossing the dam to above 70gm, at which point they are too big for the birds to feed upon (Sullivan 1991). This case suggests that it is important to understand quite detailed components of the breeding and feeding strategies of some species before embarking on habitat manipulation.

5.10.2 Freshwater and natterjack toad

The rare natterjack toad is at the limits of its range in western Europe. The grazed upper saltmarshes of northwest England and southwest Scotland support large populations of this species where it breeds preferentially in freshwater pools on the upper edge of the marsh. Occasional inundation, by spring high tides, helps keep them free from predators and the spawn of competing common frog (*Rana temporaria*) and common toad (*Bufo bufo*). Once the pools become fresh again the natterjack toad spawns and, because of its extended breeding season, has a competitive edge over the other species. Although this is a risky strategy, as further tidal inundation can occur and destroy the eggs, in years where conditions allow, good survival rates can be achieved (Banks et al. 1994).

These two examples illustrate the complex nature of some of the interactions between saltmarsh and the species which depend upon it. The conservation manager should be aware of the potential implications of any action, which as the example of the bittern shows might compromise other even more important interests.

5.11 Sea defence, sea level rise and saltmarshes

Changes in the configuration of saltmarshes are a natural part of their development. Accretion in one area may be balanced by erosion elsewhere. In the Solway Estuary, for example, alteration of the course of the main channels has caused cycles of accretion and erosion in the saltmarsh at Caerlaverock, Scotland (Marshal 1962), in the Severn Estuary, England/Wales (Allen 1992) and is a well established phenomenon for Morecambe Bay, northwest England (Gray 1972). Thus left to themselves natural systems can accommodate changes to the environment. Changes in sea level are one of the factors forcing movement either landward or seaward, depending on the way in which relative sea level is moving.

Ranwell (1972a) refers to "rising" and "falling" coastlines where the availability of sediment is important to the type of development which takes place. Tidal action and storms also contribute to natural fluctuations in the marsh. Under normal circumstances, and in the absence of human interference, there is no need for intervention whether saltmarshes are accreting or eroding. However human settlement, including saltmarsh enclosure and use for intensive arable production, results in a requirement for protection of the land from erosion and flooding.

5.11.1 Maintaining the line of defence

The biggest single conflict occurs where there are continuing new proposals for enclosure of saltmarsh. Today most major developments affecting loss of saltmarsh are associated with whole estuaries or other coastal wetlands (Chapter 10), for example tidal barrages, marinas, ports and harbours. There is a reduced pressure for new enclosures for agriculture, though proposals continue to arise in a few locations.

In areas where the saltmarsh is valued as part of a nature reserve or a first line of sea defence, direct loss of habitat may no longer be a threat. However even in these areas a major area of potential conflict still exists. As has been discussed above, much coastal land, particularly around major estuaries, has been derived from former tidal saltmarsh. Many of these areas lie below high water and there is a natural desire to prevent erosion and flooding by maintaining the existing line of defence. As has been discussed above this approach may result in further erosion of the marsh and a reduced ability of the tidal zone to accommodate changes in sea level, occurrence of storms and alteration in tidal regime. This situation may require reconsideration of policies seeking to protect the saltmarsh and the sea wall in all cases and at all sites.

5.11.2 Managed retreat

In areas where relative sea level is rising and where the natural transition between tidal waters and the land has been truncated by an artificial barrier, the outcome is the same. Saltmarshes are 'squeezed' between the development on the land and the rising sea level. Maintaining the current line of sea defence, particularly in these areas, may make the loss of saltmarsh inevitable. In turn this loss may ultimately result in the artificial sea defence itself being undermined (Figure 5.36 above). The situation on the saltmarshes of Essex and North Kent, reported above may be only an indication of what might happen elsewhere, particularly if the predicted increase in the rate of sea level rise caused by global warming is realised. It is predicted that in Essex, in southeast England, 40% of the saltmarshes may be lost over the next 60 years unless positive measures are taken to recreate them (Boorman 1992).

Rehabilitating saltmarshes is one way of protecting existing sea walls against erosion and adjacent land from flooding. In Holland and Germany a variety of methods have been used to recreate saltmarshes by the use of 'sediment fields'. Brushwood groynes and artificial drainage are used to encourage sediment deposition and saltmarsh development, which help reinforce sea defences, as well as providing opportunities for subsequent enclosure (Kamps 1962, Dijkema & Wolff 1983).

The concept of managed retreat onto areas of enclosed former saltmarsh is a relatively new concept. It has an attraction for nature conservationists because, potentially at least, it provides an opportunity for the recreation of lost tidal habitats and transitions to terrestrial vegetation (Figure 5.40). It is also a natural consequence of the ideas associated with combating the 'coastal squeeze'.

If natural habitat is being lost and the ability of the coast to accommodate changes in tidal regimes (associated with changes in sea level, storms, climate etc.) is reduced, then a logical solution is to make the coastal zone wider and more flexible. Managed retreat embodies this concept by allowing natural processes to reassert themselves. Restoration techniques have been evaluated as part of a guide to saltmarsh management prepared by the Environment Agency, which is responsible in England and Wales for sea defence (Environment Agency 1999). However, overcoming the inherent prejudices against "sea reclamation" or "depoldering" as the process has been referred to in the Netherlands (de Ruig 1998), may be more difficult than identifying successful engineering techniques.

Figure 5.40. Enclosed tidal land, formerly used for agriculture, provides opportunities for the recreation of saltmarsh when earthen seabanks decay. Note the dead trees along former field boundaries, Stour estuary, Suffolk, southeast England. Compare with an 'engineered retreat' pp. 289-96 and Figure 11.10 (Packham & Willis 1997).

6. *SPARTINA ANGLICA* - A CASE OF INVASION

In the United States of America the Federal Fish and Wildlife Service estimate that there are approximately 6,300 non-native invasive plants and animals which pose a threat to wildlife. This chapter describes one of these species which has had a profound impact on coastal tidal areas in many parts of the world. It highlights some of the questions posed for wildlife conservation managers and amenity interests.

Spartina is a widespread and often dominant plant of saltmarshes in the northern hemisphere. In North America the two native species normally encountered are *S. alterniflora* and *S. patens.* The former is a large tall grass which dominates much of the lower saltmarsh and creek banks; the latter is a fine, small plant found at higher levels of the marsh. In Europe two native species occur as perennial pioneer grasslands of mudflats, *S. maritima* and *S. densiflora.* The former is rare though widely scattered in the northwest, the latter restricted to the southern Iberian Peninsula. Whilst the native American species are important components of the saltmarsh habitat there, in Europe and many other parts of the world, it is a hybrid species *Spartina anglica* which is the most significant plant from a conservation point of view. This species may be the greatest single agent for change in saltmarshes in many temperate regions of the world.

6.1 Origins and colonisation

The American *S. alterniflora* was thought to have been brought to Southampton, in England in ships' ballast in the early 1800s. Hybridisation between this species and the British *S. maritima* sometime prior to 1870 in southern England created a cross resulting in a sterile hybrid *Spartina x townsendii* being formed. This plant, through a process involving the doubling of its chromosomes, gave rise to a fertile amphidiploid *S. anglica,* the subject of this chapter. There is a large literature relating to this species. The origin, history and spread of *Spartina anglica* is summarised by Doody (1984), and updated by Gray & Benham (1990) and Adam (1990 pp. 78-87).

6.1.1 Colonisation

The ability of *Spartina anglica* to colonise tidal mudflats through natural expansion was seen as an aid to coastal defence and land claim. As a result this invasive plant was exported from its origins on the south coast of

England to many places in Great Britain and throughout the World (Ranwell 1967). In Great Britain this led to the development of approximately 5,000ha of marsh (largely confined to the estuaries of England and Wales) representing more than 25% of the total area of saltmarsh. The saltmarsh survey of Great Britain (Burd 1989) recorded 7,000ha from records taken between 1981 and 1988. It continues to expand in northern localities such as Morecambe Bay, northwest England (Whiteside 1987) and locally in the southeast (Figure 6.41).

Figure 6.41. Spartina expansion, Blackwater Estuary, Essex, southeast England

In China, its export in the early 1960s resulted in the growth of some 36,000 ha of saltmarsh over a period of 25 years (Chung 1990). It has become well established in Australia (Boston 1981) and many other parts of northwest Europe. Its expansion continues today in Europe. In Washington State, San Francisco and Florida, USA, it also gives cause for concern.

6.1.2 The nature of *Spartina* invasion

The hybrid *Spartina anglica* is able to colonise intertidal flats and mudflats rapidly, through extensive seedling establishment in favourable years; it also spreads by the growth of detached rhizome fragments. It is very productive with a rapid growth rate (Long & Mason 1983) and produces monospecific swards as individual clones coalesce. It also has the ability to spread lower down the normal tidal range of perennial saltmarsh vegetation (Gray 1990). These characteristics can greatly accelerate the rate of saltmarsh growth.

In the Dee Estuary, on the border between Wales and England, the introduction of *Spartina* in 1928/29 appears to have considerably accelerated a process begun earlier by pioneer *Salicornia europaea* and *Puccinellia maritima*. Fishing boats were still able to tie up at Parkgate in the 1940s whilst today there is a marsh nearly 2km wide fronting the former quayside (Pye 1996, Figure 6.42).

6.2 Nature conservation implications

In Great Britain it was widely accepted that the rate and scale of colonisation by *Spartina anglica* had been detrimental to nature conservation interests. Implicit in the review of its occurrence and impact on tidal flats and saltmarshes (Doody 1984) was that it was an invasive species with serious implications for nature conservation. In particular two issues were considered important: the impact on native plant communities and the effect on the fauna, notably wintering waterfowl, of tidal mudflats.

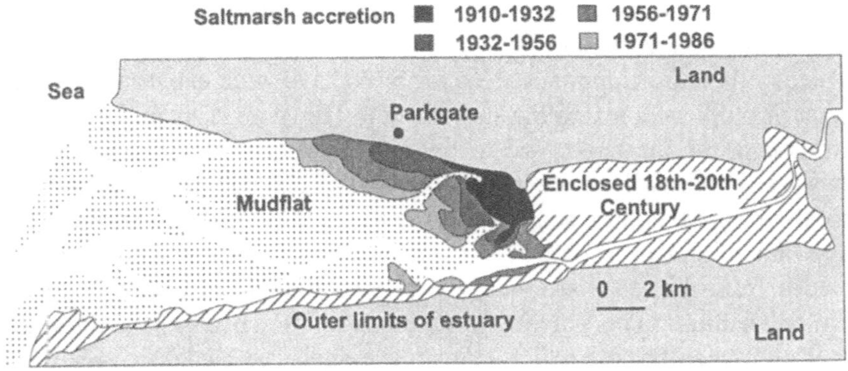

Figure 6.42. Spartina colonisation alongside the former quayside at Parkgate on the northern shore of the Dee estuary, Cheshire, England

6.2.1 Plant species effects

Botanically, the effect of *Spartina* invasion on the native flora seems clear. As it replaces the natural pioneer plants, including *Puccinellia maritima*, *Salicornia europaea* and *Suaeda maritimus* (in Europe), the course of succession is altered. Once established, two forms of development have been seen to occur. Firstly on the grazed marshes in northwest England *Spartina* is replaced by swards of *Puccinellia maritima*. Experience from the Ribble Marshes nature reserve indicate that this succession can take place over a relatively large area in one season. The earlier maturing *Puccinellia maritima* is able to out-compete *Spartina* at higher marsh levels. Secondly on ungrazed sites *Spartina* may be replaced by stands of *Phragmites australis* or *Scirpus maritima* at the upper margins of the saltmarsh. The species composition of the communities which result are generally impoverished by comparison with those developing in the absence of *Spartina anglica*.

Early reports of its invasion of the mud flats refer to the replacement of *Zostera* beds (Oliver 1925, Chapman 1959). Ranwell & Downing (1960) consider the decline in *Zostera* to be caused by a 'wasting disease', which was further aggravated by the rapid spread of *Spartina*. The native *Spartina maritima* has also declined in its range since the advent of *S. anglica*; however, there is no known direct causal link. More recent reviews suggest that the decline in *S. maritima* is continuing at least on the south coast of England (Gray et al. 1999).

6.2.2 Ornithological effects

Tidal sand and mud flats support large populations of invertebrate animals which are prey for wintering wildfowl and wader species, whose numbers may attain international significance in many estuaries (Chapter 10). Typically the most important areas for birds have wide expanses of tidal flats appearing to have a stable low water mark. However rises in sea level cause a truncation of the shore and a lowering of its profile (Pethick 1996). In estuaries, particularly those with deep river channels and steeply shelving shorelines, there is little scope for further accretion of intertidal flats, even if there are abundant sediments. This effectively means that, as *Spartina* invades from the landward side, there is an overall loss of open sand and mudflat habitat. This gives most concern for ornithological interests, especially when the expansion takes place against a backdrop of land-claim which has already diminished the available intertidal land.

Potential problems were identified early this century when *Spartina* was being promoted as an aid to stabilisation and reclamation. Stapf (1913), for example, cites 'interesting economic effects in the fauna', including the loss of large molluscs which were collected for food. At that time impacts on the bird populations were not a consideration. However, evidence that *Spartina* was having a significant effect on wader populations began to mount from the mid 1980s. Goss-Custard & Moser (1990) discuss the relationship between the decline in dunlin numbers and the spread of *Spartina* in several British estuaries. At one site, Wader Bay in the Dyfi estuary, Wales, *Spartina* invaded only 10% of the intertidal flats. However, this invasion was at the expense of some of the most favoured wader feeding sites leading to a disproportionate adverse effect on feeding dunlin (Davis & Moss 1984). The loss of intertidal feeding areas and the steepening of the beach profile were also thought to be a potential threat to bird populations at Lindisfarne National Nature Reserve (Millard & Evans 1984). More recent studies at the same site suggest that changes in sediment patterns and *S. anglica* encroachment have caused a reduction in *Zostera* coverage with potential adverse effects on grazing wildfowl (Percival et al. 1998).

6.3 *Spartina* in the USA

"Imagine Puget Sound without its gravel beaches, shellfish beds and waterfowl. Imagine Puget Sound without its eel grass beds, salmon stocks and quiet shallow lagoons. This disturbing picture could become reality if *Spartina*, a non-native grass, is allowed to spread unchecked along Puget Sound's shoreline. It has already gained a foothold in bays and lagoons as far north as Padilla bay and as far south as Vashon Island." (Extract from Washington Water Trails Association Web Site @ http://www.eskimo.net/~wwta/environ/spartina_where.htm).

Spartina was introduced to Puget sound's intertidal lands approximately 35 years ago, following an earlier attempt some 65 years previously. The species has completely changed the character of inter-tidal mud flats, crowding out native plant communities and destroying critical migratory waterfowl and shorebird feeding areas. Today it has spread throughout the area (Figure 6.43) and is of major conservation concern.

Figure 6.43. Spartina alterniflora distribution in Washington State

It seems that the species, *S. alterniflora*, native on the east coast of America, has along with other species of *Spartina*, found a niche on the northwest coast particularly conducive to its survival. In Washington State two international conferences have been convened, specially to consider the issues surrounding the control of the species (Aberle 1990), the most recent in 1997 (Patten 1997).

6.4 Methods of management

The potential impact on nature conservation interests has been the main driving force for considering and using control measures along with amenity concerns at a smaller number of locations. As with most species rapidly invading open habitats, once established *Spartina* is difficult to eradicate. Problems can be avoided if the species is controlled when it first appears. In this context hand-weeding seedlings can be effective, though time consuming. Successful attempts have used voluntary labour, as for example is the case in Puget Sound, through its *Spartina* Watch volunteers scheme. Here people are asked to survey agreed sections of shoreline; volunteers also have the opportunity to work with others to remove *Spartina* from the beaches. Methods of control include hand-pulling seedlings and digging up small clumps or individual plants of *Spartina* by the roots.

In England attempts were made to transplant rhizomes of *Spartina anglica* suffering from 'die-back' from the east coast (Norfolk) to the northeast coast, in order to see if some pathogen might be implicated. This was not successful. Herbicide control has proved to be the most successful approach. Several compounds have been tried, of which dalapon has been shown to be effective (Evans 1984, Way 1987). Difficulties in obtaining dalapon resulted in trials of glyphosate in the United Kingdom but these have not developed as licensing problems restricted their use.

Biological control has also been advocated (Wecker 1998) though specific trials had not been attempted. More recently Washington State lawmakers made funding available (1998) for trials to test biological control methods on *Spartina*. There are four insects in Northern California that are *Spartina*-specific and a leaf hopper *Prokelisia marginata* eats only two specific taxa including *Spartina alterniflora*. Of the other methods tried, recent experiments (1996) at Lindisfarne National Nature Reserve in northeast England have shown encouraging responses to burying of the plants by ploughing and rotavation.

A summary of the effectiveness of a variety of mechanical methods is presented in Table 6.15, partly derived from Way (1987).

Table 6.15. Spartina control - a summary of mechanical treatments

Method	Effectiveness	Advantages & disadvantages
Digging	Partially & on a small scale (mainly seedlings)	Labour intensive, & costly on a large scale
Dyking and inundation	Partially effective in preventing spread	Costly & damaging to other saltmarsh communities
Bulldozing (removal of surface)	Ineffective	Potential damage to mud surface
Rotavating & Harrowing	Counter-productive	Greater propagation from broken rhizomes
Burying (ploughing & rotavation)	Effective if plants are covered	Difficulties of access
Crushing	Partially effective	Requires repeat treatment, vehicular access difficulties
Burning	Ineffective	Impractical

Grazing	Prevents seedling production, no reduction in clumps	Increases shoot density
Mowing	Prevents seedling production, no reduction in clumps	Can increase shoot density
Covering (black plastic)	Partly effective on a small scale	Difficult to keep plastic in place

6.4.1 Lessons from an amenity beach

Nature conservation interests are not the only justification for considering *Spartina* management. Amenity concerns have also led to control being attempted both in the United Kingdom and North America. In the UK along the Sefton Coast in northwest England, a change from a sandy to a muddy shoreline prompted the local authority to undertake *Spartina* control, it having been implicated in the change. The effectiveness of this method of control and the implications for the nature conservation interest within the area has been reviewed (Bennett 1996). Although the expansion of the marsh has been partially controlled, the work is expensive and the marsh continues to expand.

It is important to understand the precise role of *Spartina* in initiating the change and this is far from clear. Sefton Coast lies on the outer extension of the Ribble Estuary where there is a major offshore sand extraction point. At the same time past land claim for agriculture (Figure 1.3) and the building of a river training wall have influenced the tidal and sedimentary regime along the coast, causing changes within the estuary. If the outer margins of the estuary are predisposed to siltation as a result, then *Spartina* expansion may only be a reflection of the changing situation rather than a cause. Thus its control may be effective only in the short term and, if *Spartina* does not invade, other saltmarsh plants may well do so. This is obviously important when considering the method and efficacy of control measures to be adopted.

6.5 *Spartina* friend or foe?

Given all the problems identified above it is easy to conclude that *Spartina anglica* is a major threat to nature conservation interests, especially in estuaries. In the early days of its introduction in Great Britain (and elsewhere) 'stabilisation, coast protection and reclamation' were the aims. This increased the opportunities for saltmarsh enclosure and with it adverse consequences for conservation interests (Chapter 10). From this perspective the further expansion onto tidal flats was considered to be at the expense of waterfowl feeding habitat. However, establishing a causal link between its spread and adverse impacts on nature conservation and amenity interests is not entirely conclusive (Thompson 1990). In this context it is important to look more carefully at the reasons for its control since this can be costly and may be ineffective.

6.5.1 Benefits

High primary productivity is a characteristic of *Spartina* spp. and can make a significant contribution to overall estuarine productivity (Long & Mason 1983). In its turn this may help support the large numbers of invertebrate herbivores of the mudflats, their predators and the birds which feed upon them. It may also be important as part of the fish-nursery role played by estuaries. The expanding edge of an actively growing *Spartina* marsh in north Wales, for example, has been shown to be an important nursery area for sea bass (*Dicentrarchus labrax*) (Kelley 1986).

Quite extensive areas of saltmarsh derived from *Spartina* are also used by grazing animals. It is reasonably palatable, particularly for the older breeds of sheep, and there are a number of sites in the United Kingdom where it is grazed, e.g. Bridgwater Bay in Somerset and Traeth Bach, Gwynedd in Wales. However, it is the flat lawns dominated by *Puccinellia maritima* and *Festuca rubra* which provide the most palatable grazing, and extensive use is made of these in the north and west for cattle and sheep. Such marshes may develop from *Spartina* marsh particularly when the latter are grazed, as in Bridgwater Bay, the Ribble Estuary and in Morecambe Bay as discussed above.

In China the extensive swards have been claimed for agricultural use, for pasture and animal fodder, coastal stabilisation, fuel, paper making and scientific study (Chung 1990). As a saltmarsh it also has benefits for sea defence which may result in direct conflict between nature conservation interest and coastal protection authorities, when its control is promoted.

6.5.2 Conservation issues

In summary, *Spartina* may be regarded as harmful to nature conservation interests because it:

- invades intertidal flats which are rich in invertebrates and are the feeding grounds of large numbers of over-wintering waders and wildfowl;
- replaces a potentially more diverse pioneer plant community;
- produces dense, monospecific swards which change the course and pace of succession and are replaced, in ungrazed areas, by tall communities equally poor in species;
- helps promote the reclamation of land for agriculture, thus destroying species-rich, high-level saltmarsh.

On the plus side it is likely to help create saltmarsh and in areas where relative sea-level is rising it may help offset the losses resulting from erosion, such as those reported from southeast England (Chapter 5).

6.5.3 'Die-back'

There is another characteristic of the species which should be taken into account when considering control of mature stands. This is a phenomenon called 'die-back' which has been observed in the dense swards of *Spartina anglica* over many years in southeast England where extensive areas of marsh have been lost. The situation has been described for a number of sites on the south coast of England - see for example Langstone Harbour (Figure 6.44). The reasons for this loss are not well understood, with a variety of theories being put forward, including the development of toxic conditions associated with the physical nature of the sediments (Adam 1990).

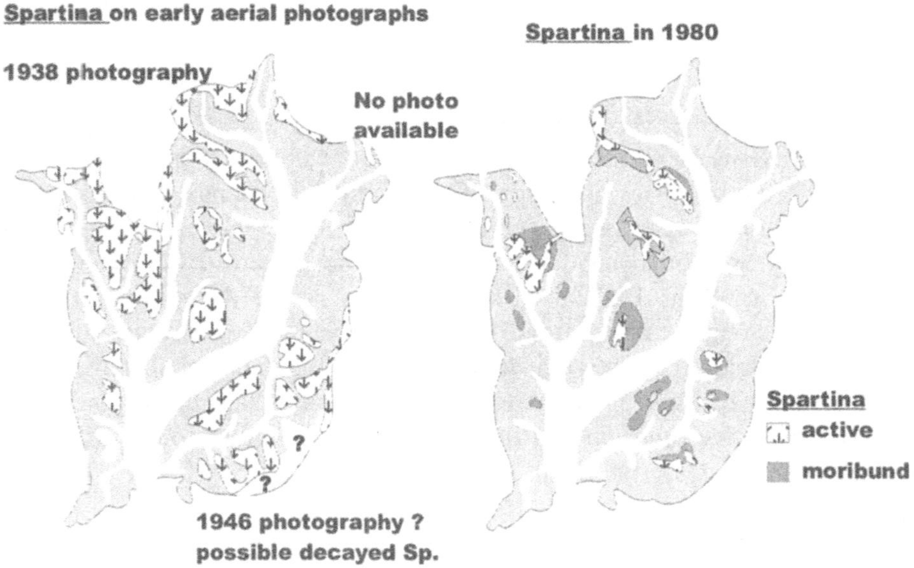

Figure 6.44. Loss of *Spartina* marsh through 'die-back', Langstone Harbour, Hampshire, England (Haynes 1984)

The situation throughout the UK further strengthens the importance of reviewing the status of *Spartina* and the reasons for it before embarking on control measures. A look at figures published for its growth and decay show a distinct geographical trend (Table 6.16). This suggests that the east coast has seen the greatest decrease, followed by the south coast whilst in the northwest the species is still colonising new sites and expanding in those where it existed already. The timing of these changes may be significant as may be the change in relative sea level. Much of the 'die-back' had already occurred on the south coast in the most mature communities (Figure 6.45), hence the rather lower figure for this geographical area. The situation on the east coast is less clear as there have been fewer studies, but expansion probably came later (Doody 1984). The evidence from Robinson & Pringle (1987) and Gray & Benham (1990) also show the later expansion in the northwest.

Table 6.16. Area of *Spartina anglica* as given for three areas in Britain (Charman 1990)

	East coast	South coast	West coast	Total
Hubbard & Stebbings (1967)				
Area of *Spartina* (ha)	6,568	3,326	2,312	12,205
Number of sites	27	24	35	86
Updated figures				
Area of *Spartina* (ha)	3,655	2,951	3,248	9,854
Number of sites	27	29	55	111
New sites ⇔ old sites	(+1 -1)	(+6 -1)	(+20)	(+27 -2)
Change	44% reduction	11% reduction	40% increase	19% reduction

The reasons for these differences have not been investigated to any great degree. However, a combination of the age of the stands themselves and relative sea level rise (greatest in the east and south, stable or falling in the west) may be influencing the patterns. The development of 'rotten spots' (Packham & Willis 1997, p. 101) in marshes on the Atlantic coast of the USA are thought to be caused by waterlogging (Mendessohn & McKee 1988). It is interesting to speculate if this effect could also be linked to rising sea levels. The development of anoxic conditions, causing decline in the vigour of native *Spartina* and death suggest that similar forces may be at work in both the UK and North America.

6.6 Is there a case for control?

It seems clear that where *Spartina* is rapidly invading new territory it does pose a threat to nature conservation interests. Research in the United States suggests that under these circumstances its control would seem to be desirable. However, since the major interest in the species in the 1980s in the UK, little research work has been done on long term changes in status and the implications for coastal conservation interests. For example what are the likely effects of global warming and a rise in sea level? Will *Spartina* marshes decline as they age and are swamped by more frequent tidal inundation? If so, are our present attempts to control its spread really justified? *Spartina* remains a fascinating genus which would repay renewed interest in understanding its biological and ecological effects.

In Europe the legislation associated with the European Union Habitats and Species Directive includes communities of *Spartina* as follows:

"15.2 *Spartina* swards (*Spartinion maritimae*) Perennial pioneer grasslands of coastal salt muds, formed by *Spartina* or similar grasses. When selecting sites, preference should be given to those areas supporting rare or local species: *Spartina maritima*, *S. alterniflora* and *S. densiflora*."

S. anglica is also included and in so far as it forms a part of the variation of saltmarsh development in so many estuaries its control should be given very careful consideration. It could be argued that the species is at a very early stage in its evolution. Given time, it might be considered to be an integral part of the natural saltmarsh landscape as is one of its parents (*S. alterniflora*) in eastern USA. Whatever the balance between the needs of bird life and the flora of the native saltmarsh plant communities, any decision to undertake control must look at the wider implications of management. In this context it will be important to understand the way in which the whole sedimentary system operates. It may be that *Spartina* invasion is a symptom of change not a cause.

Figure 6.45. 'Die-back' of mature marsh in the Solent, Hampshire, England

7. SAND DUNE

7.1 Introduction and scope

Sand dunes border long stretches of the World's coastline and are best developed in temperate and arid zones (Bird 1984). They are formed when sandy shores dry out and sand grains are blown inland. Accumulations a few centimetres to 40m or more thick are formed by the combined action of wind and the stabilising effects of vegetation. Whilst most dunes show contemporary contact with the sea, on rising shorelines or where sand accumulation is extensive this contact may be discontinuous or absent. In Queensland, Australia, dunes exceeding 275m in height have developed through episodic migration (Bird 1984) and in Scandinavia land uplift has resulted in dunes occurring today far removed from the coast (Aartolahti 1973, Heikkinen & Tikkanen 1987). This chapter is concerned largely with dunes which are derived from sediments driven onshore by the action of the sea and subsequently moved inland by wind. Throughout reference is made to geomorphological issues though this is not dealt with in detail, but see Pye & Tsoar (1990), Nordstrom et al. (1990) and Bakker et al. (1990). The extensive barrier island coast of the eastern USA is mentioned only in relation to considering sea defence issues.

From a nature conservation point of view sand dunes, along with sea cliffs (Chapter 3) and saltmarshes (Chapter 5), are often considered to be amongst the most natural habitats. However, as with other habitats covered here, this naturalness belies a long history of human intervention. This ranges from the indirect effects of deforestation in upland areas, resulting in the delivery of large quantities of sand to the coast which help to form micro-tidal deltas (Chapter 10), to changes in vegetation patterns brought about by the use of the dunes for grazing. The famous Neolithic village of Skara Brae in Orkney testifies to human occupation dating back some 5,000 years between 5,100 and 4,450 BP (Ritchie & Ritchie 1978). In South Africa the destabilising effects of human activity have been traced back to Stone-age man (Bate & Dobkins 1992). Historically, far from being empty places where sand movement precluded exploitation, they were often areas where settlement took place. In Great Britain, Denmark and Holland dunes became important areas for supplying rabbits (*Oryctolagus cuniculus*), and in Holland people gathered firewood and winter feed, cut sods for houses and *Ammophila arenaria* for thatch and grazing animals (Wallage-Drees 1988).

From this it appears many sand dunes were busy places and their structure and vegetation today reflect a long history of human intervention and exploitation. The underlying principles of dune development, including species interactions, community dynamics and physical conditions are dealt with in chapter 6 of Packham & Willis (1997).

7.1.1 Habitat definition

Sand dunes develop wherever there is a suitable supply of sediment within the size range 0.2-2.0 mm. The form of the dune depends on the amount of available sediment and the ease with which it can be moved by the wind. Wet sand usually remains *in situ*. Thus sand dunes may be defined as:

"areas where wind blown sand is deposited inland from a wide beach which dries out periodically".

Mobile sand grains come to rest in areas where wind velocities are reduced in the lee of obstacles on the beach, open dune ridges or more typically, in temperate regions, around specialist plants. As the dune develops, other factors come into play and under the influence of grazing the vegetation may range from calcareous grassland to acid heathland. The extent of the latter depends on the original calcium content of the sand and the length of time during which leaching of the surface layers takes place. Elsewhere, especially in the warmer south, a variety of scrub and woodland communities develop. Dune slacks are also an important component occurring at or near the dune water table, which may be strongly domed. They are rich in species, particularly when associated with calcareous sand.

7.1.2 Habitat type

Sand availability and mobility are key components in sand dune development and coastal sand dunes can be categorised according to their geomorphological structure. There are 6 main types grouped according to the underlying topography and climate in which they develop (Figure 7.46 & 7.47 after Ranwell & Boar 1986).

These are described as follows:

- **Bay dunes** are the most common and are found associated with indented coastlines often between headlands;
- **Hindshore dunes** may become very large as both dominant and prevailing winds blow sand inland in sometimes massive waves. On especially exposed sites coastal cliffs may be covered with a veneer of sand blown up over the cliff to create 'climbing dunes'. The machairs of the west of Ireland and Scotland are also a special kind of hindshore dune which occurs as a flat sandy plain;

- **Barrier islands** tend to occur in exposed locations and may be built on shingle bars; examples include the Wadden islands of Denmark, Germany and Holland and the North Norfolk coast in eastern England;
- **Ness dunes**, including cuspate forelands, occur where prevailing and dominant winds are in opposition. Sometimes progradation may be quite rapid and a series of low dune ridges occurs interspersed with wet hollows. These are also found in areas where sea level is rising relative to the land as in southeast Norway, northeast Scotland and Northern Ireland;
- **Spit dunes** are formed when sediment is deposited at the mouth of an estuary. They occur in a variety of forms depending on the action of waves (longshore drift) and the force of the river flows to the sea. These include the dune forms associated with deltas in micro-tidal areas.

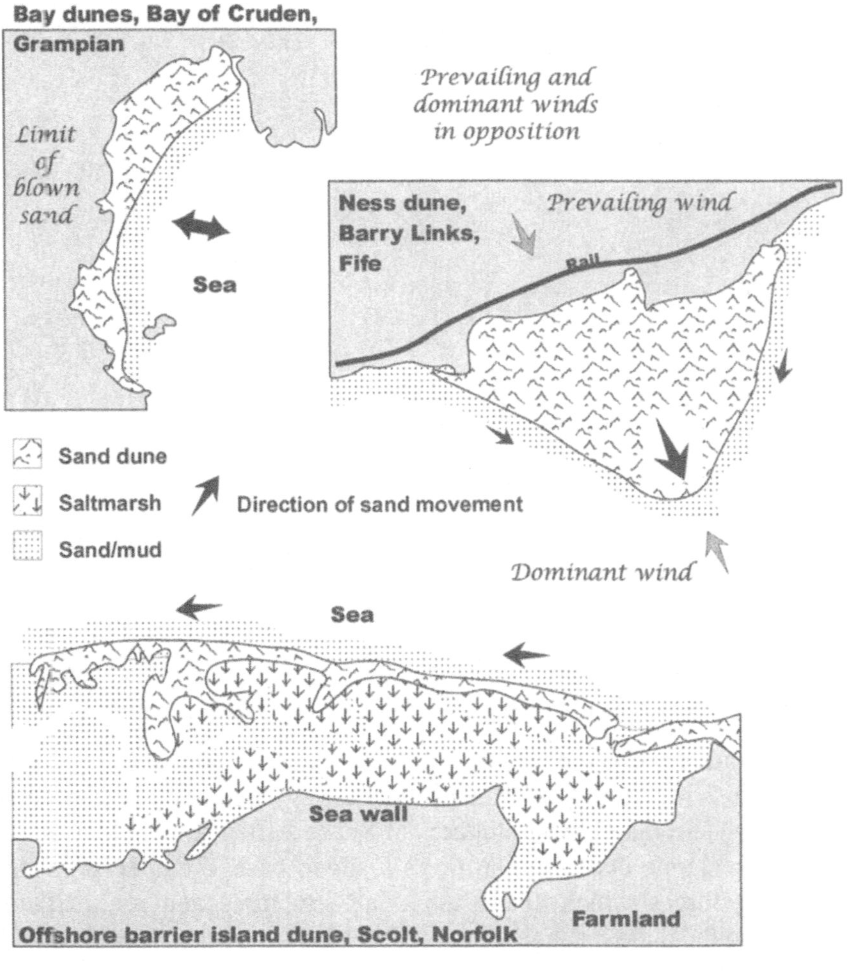

Figure 7.46. Physical sand dune types taken from examples in Great Britain

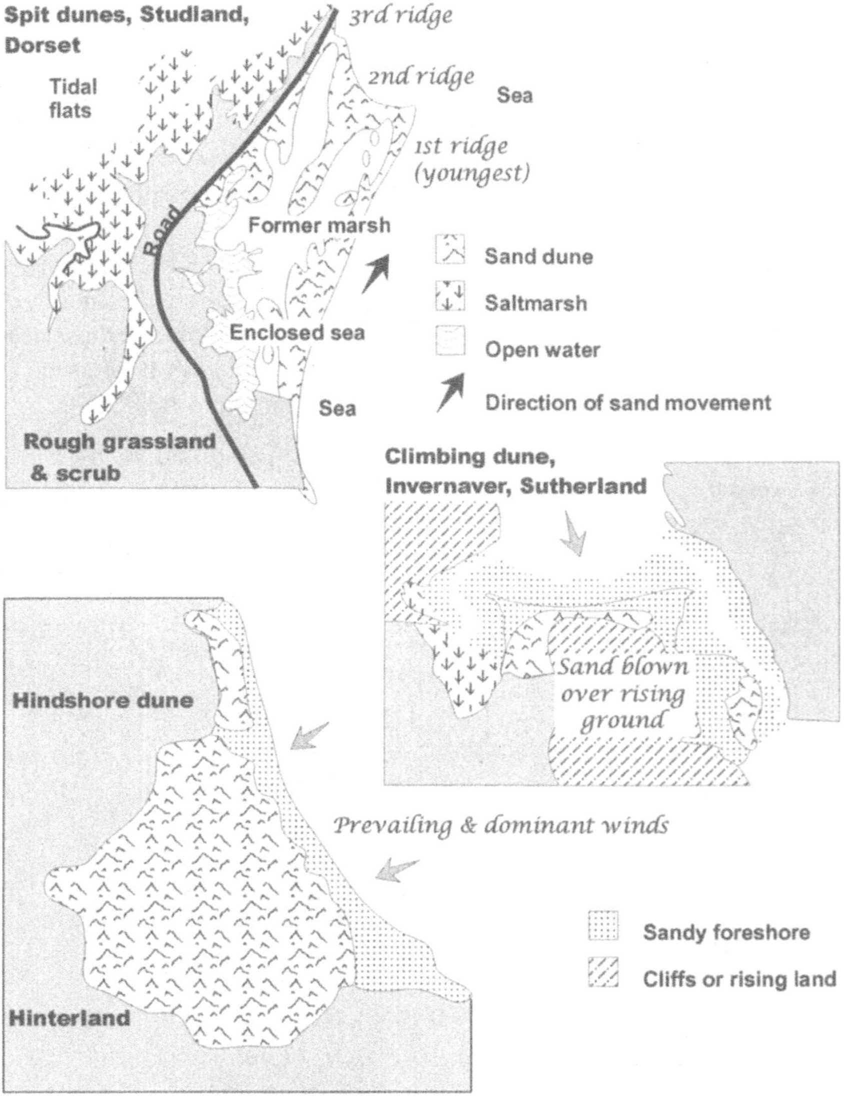

Figure 7.47. Physical sand dune types taken from examples in Great Britain, continued

Sometimes an offshore island is connected to the mainland by a dune spit or barrier to form a tombola. Major hindshore systems may cross headlands, the so called 'headland bypass dunes' of South Africa (McLachlan & Burns 1992). This simple classification gives a sound basis for understanding the dominating forces which shape the dune structures and are particularly important to the dune manager.

7.2 Sand dune distribution

Sand dunes occur extensively around the temperate coasts of the world. The combination of sediment availability, strong winds and vegetation facilitate their development. Sand dunes are less frequent in the tropics due to the more luxuriant vegetation, low wind velocities and damp sand conditions (Pethick 1984).

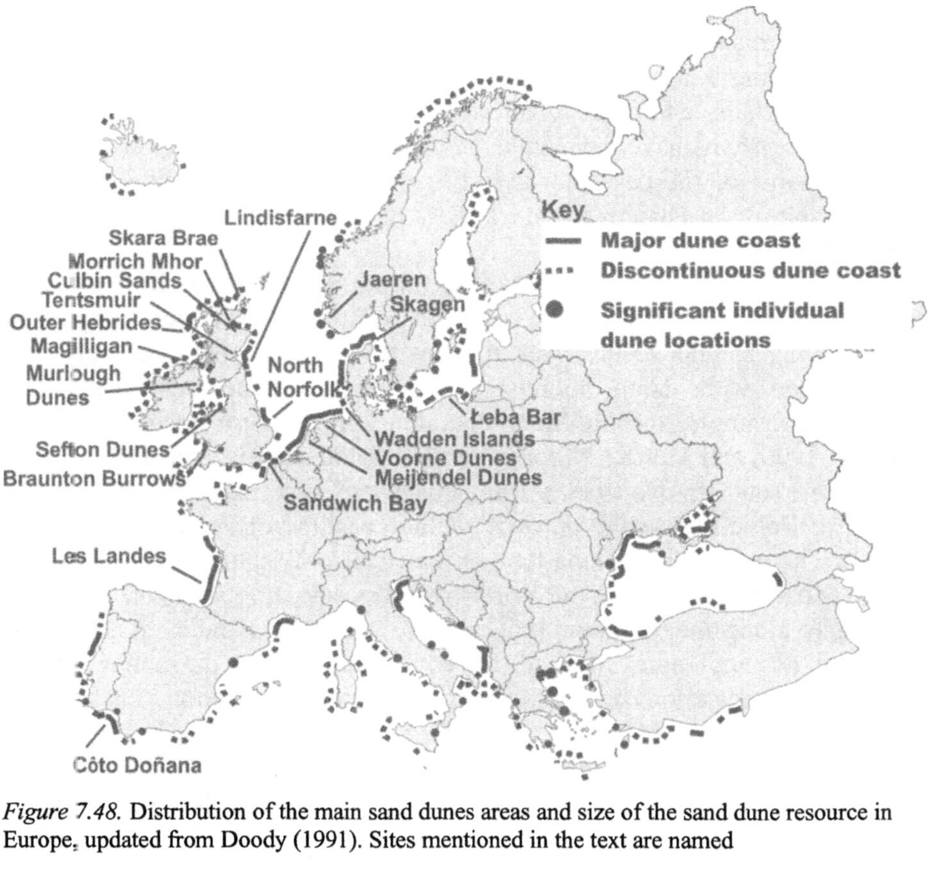

Figure 7.48. Distribution of the main sand dunes areas and size of the sand dune resource in Europe, updated from Doody (1991). Sites mentioned in the text are named

7.2.1 Habitat distribution in Europe

Throughout Europe and in other temperate regions, sand dune development is intimately bound up with the stabilising effects of vegetation. The overall distribution and size of the dunes is a reflection of the key influences of sediment availability, strength and direction of the prevailing and dominant winds and the physical nature of the coast. Figure 7.48 (above) shows the distribution and overall area of European dunes derived from a survey based on published references and information provided by country specialists

(Doody 1991). Other sources include national surveys of Denmark (Brandt & Christensen 1994), Great Britain (Dargie 1993, 1995, Radley 1994) and Ireland (Quigley 1991).

Atlantic, North Sea and the Baltic

Much of the coastline of the northeast Atlantic coast is formed from ancient rocks resistant to erosion. The absence of sedimentary material for the development of sand dunes mean that there are relatively large numbers of small sites, many of which are associated with embayments. Exceptionally, on west-facing coasts where prevailing westerly winds are reinforced by dominant winds, and large quantities of shell sand are present, massive hindshore systems have developed. In the Outer Hebrides in Scotland these include some of the best and largest examples of the extensive cultivated sandy plain or machair made up largely of shell-sand, also present in the west of Ireland.

The northern parts of the Celtic Seas, North Sea and the Baltic Sea dunes develop on coastlines which are rising relative to sea level. Here prograding ridges lying parallel to the coast form as a sequence. They are sometimes interspersed with damp hollows in which rich dune slack vegetation develops. Examples include Magilligan dunes in Northern Ireland (Carter & Wilson 1990) and Morrich Mhor in northern Scotland (Hansom 1999).

In the southern North Sea sand dunes are very extensive and in the southern Baltic they make up 80% of the coastline of Poland (Piotrowska 1989). The predominant dune type is represented by barrier islands (e.g. in the Wadden Sea and along the North Norfolk coast, England) and sand bars and spits along the southern Baltic coast, which lie parallel to the coast. Massive accumulations of sand forced onshore under the action of the prevailing winds also occur in northern Denmark and Holland.

Dunes or dune remnants are found along the coastline from western Ireland, southwest Britain and northern France, Spain and Portugal. In this region the coast has a predominantly cliffed nature and the availability of suitable sedimentary material is restricted. Here, as with the cliffed landscapes further north, dunes tend to be smaller and develop in sheltered embayments. Along exposed west-facing coasts of France and Portugal abundant material from fluvial sources, in combination with the strong erosive action and long shore drift, result in a large sediment transport system. In these areas dunes may stretch many kilometres inland. In France, for example, the area known as 'Les Landes' has a special significance, and is one of the most extensive duneland areas in Europe stretching some 230km along the coast and in some places reaching up to 70km inland (Barrère 1992).

Mediterranean and Black Sea

In Spain and southern Portugal the dune systems include barrier islands and spits. In Andalusía, river transport has helped to create one the most important dune systems in Europe. This lies within the Côto Doñana National Park and consists of beaches, foredunes, mobile dunes up to 30m high and stabilised dunes which enclose a major wetland.

In Italy dunes probably originated during the thermal optimum after the last glaciation (8,000 years ago). They have since been broken by rivers and their subsequent development has been a product of natural erosive forces and use by man.

There are many places in this region where sand dunes cannot develop because the hills or mountains outcrop as rocky shorelines plunging steeply into the sea. Along the coast of Croatia, southern Albania and many of the Greek islands, dunes are virtually absent. Where they do occur they tend either to occupy a narrow fringe bordering flat areas of land or exceptionally form extensive dunes up to 10m height, as in Western Peloponnisos, in Greece. In a few areas the dunes may reach a height of 20-30m.

The extensive dunes of Turkey and on the northern shores of Albania are different. Almost all are formed in the immediate vicinity of rivers, occurring as significant elements within large delta systems. They may often form a barrier to the sea and enclose a series of lagoons. Their maximum height is about 80m and they have a maximum width of up to 4.3km as, for example, in southwest Turkey. Many different dune forms are present, e.g. huge beach plains with embryo dunes, parabolic dunes, blowouts, dune slacks, lakes, secondary barchans and dune fields. In the Mediterranean systems the calcium carbonate content is very high, while siliceous sands prevail along the Black Sea coast.

7.3 Nature conservation value

The high nature conservation value of sand dunes is derived from the character of the dunes themselves (their geomorphology), the diversity of habitat and vegetation and the rare and specialist plants and animals which live there. These interests reflect both temporal and spatial variation and inter-relationships with other coastal habitats, notably saltmarshes, shingle and lagoons, with which they are often intimately associated. The value of sand dunes for nature conservation purposes is usually assessed by reference to its vegetation. The European Union Habitats and Species Directive, for example, identifies plant communities which form the basis for assessment (European Commission 1999c, Table 7.17).

Their variability, together with the dynamic and open nature of the duneland, also provides suitable habitat for a wide variety of specialist invertebrates and rare amphibians and reptiles. Today a climax vegetation of natural dune forests is rare. Usually in northwest Europe high rainfall, combined with the long history of grazing by domestic stock or use as rabbit warrens, has resulted in stable dunes which are often dominated by species-rich grassland or heathland.

Table 7.17. Coastal habitats of Community interest (sand dunes) forming the basis for the selection of Special Areas for Conservation (SAC's). From the Interpretation Manual of European Union Habitats, European Commission, October 1999. *Priority Habitats, types

Codes	Directive Name
Atlantic, North Sea and Baltic	
16.211	Embryonic shifting dunes
16.212	Shifting dunes with *Ammophila arenaria* (white dunes)
16.1222, 16.132/3, 16.2133	Boreal Baltic sandy beaches with perennial vegetation
16.221-16.227*	Fixed dunes with herbaceous vegetation (grey dune)
16.23*	Decalcified dunes with *Empetrum nigrum*
16.24*	Eu-atlantic decalcified fixed dunes (*Calluno-Ulicetea*)
16.251	Dunes with *Hippophae rhamnoides*
16.26	Dunes with *Salix arenaria* ssp. *argentea (Salicion arenariea)*
16.29	Wooded dunes of the Atlantic, Continental & Boreal region
16.31-16.35	Humid dune slacks
1A	Machairs * in Ireland only
Mediterranean coast	
16.223	*Crucianellion maritimae* fixed beach dunes
16.224	Dunes with *Euphorbia terracina*
16.228	*Malcolmietalia* dune grasslands
16.229	*Brachypodietalia* dune grasslands with annuals
16.27 & 64.613*	Coastal dunes with *Juniperus* spp.
16.28	*Cisto-Lavanduletalia* dune sclerophyllous scrubs
16.29 x 42.8*	Wooded dunes with *Pinus pinea* and/or *Pinus pinaster*
Continental dunes, old and decalcified	
64.1 x 31.223	Dry sand heaths with *Calluna* and *Genista*
64.1 x 31.227	Dry sand heaths with *Calluna* and *Empetrum nigrum*
(64.11 or 64.12) x 35.2	Inland dunes with open *Corynephorus* and *Agrostis* grasslands

7.3.1 The importance of succession

The vegetation succession is often described in a sequence from open mobile dunes to more stable types as the growing vegetation traps sand. The zonation which develops reflects the physical characteristics affecting sand transport rates and deposition (Willetts 1989) acting in concert with the vegetation and other biological factors (Willis 1989, Packham & Willis 1997 pp. 153-169). Although the sequence rarely occurs in a straightforward progression from one stage to the next, the complexity of the spatial and temporal variation imparts a diversity not found in more stable habitats such as woodland, permanent grassland or heath.

The development of foredunes is directly related to the ability of plants such as *Elytrigia juncea* and *Ammophila arenaria* to withstand burial by sand together with the other stresses, such as water availability, in this inhospitable environment (Ranwell 1972a). Similar species occupy the same niche in America, where *Ammophila breviligulata* replaces the ubiquitous *Ammophila arenaria* of Europe (Figure 7.49).

Figure7.49. Ammophila dunes on an exposed shoreline in northeast Scotland

Once the main body of the dune is formed other processes come into play and initiate a sequence from mobile foredunes and 'yellow dunes' to 'grey' semi-fixed dunes with grassland, heath, scrub and woodland (Chapman 1964). The original calcium content of the sand, and the age of the dune soil (including the degree of leaching) helps to determine whether it is calcareous grassland or heathland. This development rarely occurs as a straight forward succession. Blowouts occur with or without the intervention of man and can be the precursors of dune slacks (Ranwell 1972a). Similarly the reprofiling of dune ridges under the influence of changing wind patterns brings an infinitely variable topography, the origins of which may be difficult or impossible to unravel. The nature of the succession is shown by reference to two areas: the exposed macro/meso-tidal areas of northwest Europe and the sheltered dunes of micro-tidal shores of the Mediterranean.

Northwest Europe

Most of the sand dunes described for northwest Europe have a sequence of vegetation types which potentially includes all the more important successional communities from strandline (driftline) to yellow and grey dune, dune pasture, heath and scrub. In areas where beach erosion is occurring some of the early stages of succession may be absent with the sand dune forming a cliff above the beach. Alternatively yellow dune vegetation can occur in blow-outs within the body of the dune.

- **Foredune (yellow dune).** The early stages of succession where dune mobility is of over-riding importance are dominated by *Ammophila arenaria* and *Elytrigia juncea*. In parts of northern Britain and southern Scandinavia they may occur together; however *Ammophila arenaria* tends to be absent further north.
- **Dune grassland.** As a more stable form of dune develops, *Ammophila arenaria* becomes less frequent. With decreasing inputs of sand plant growth diminishes, new shoots are less frequent and the individual culms become more sparsely dispersed. This process is accelerated under domestic grazing regimes where *Festuca rubra* and other grasses help form the typical species-rich calcareous dune grassland when the original sand grains have a high calcium carbonate content. In the north, *Primula scotica*, *Dryas octopetala* and *Juniperus communis* are important locally and, in the south and west, *Anacamptis pyramidalis* is amongst several orchid species which find the warm open soils especially conducive to their survival.
- **Dune heath.** Acid dune grassland (grey dune) dominated by *Deschampsia flexuosa* and *Festuca ovina* or dune heath (brown dune) with *Empetrum nigrum* and *Calluna vulgaris* develops on dunes where the calcium carbonate content of the soil is absent or low. This may be due to a high silica content and a corresponding lack of shell fragments in the original dune sand, or from a long period of leaching of the soil surface in older dunes. *Corynephorus canescens* and *Carex arenaria* with abundant mosses and lichens are often characteristic species. (Note the first of these species is found as far south as Portugal.)
- **Dune slack.** Dune slacks occur where the water table is near the surface of the sand. They are found on many dunes and are often rich in species, particularly when associated with calcareous sand. Species include a number of rarities such as *Liparis loeselii.* Some slacks may become dominated by willows (*Salix* spp.) together with *Phragmites australis* marsh in the wetter areas.

- **Dune woodland**. Because of the extent of grazing very few dunes, if any, exhibit primary woodland. However secondary *Betula pendula* for example in Scotland and Holland, and *Pinus sylvestris* in Finland still occur today. Prior to this, open vegetation was maintained by the grazing by domestic animals.

Mediterranean coast

The further south the dunes are formed, the more the southern elements of the flora appear. Typically in the Mediterranean an open vegetation occurs above the beach where *Pancratium maritimum* is a common species. There is a less obvious successional development, though narrow zones of pioneer vegetation dominated by *Elytrigia juncea* and/or *Ammophila arenaria* are present. This grades into a grassland and scrub (Figure 7.50) including garigue (an open dwarf shrub community about 60cm and rarely >1m high) and maquis (a dense shrub community >1m high) typical of grazed inland areas (Polunin & Walters 1985).

Figure 7.50. Open dune grassland with *Matthiola sinuata*, Mediterranean coast of Spain

- **Yellow dune**. The typical sequence of zonation shown by Atlantic dunes is less obvious on the Mediterranean coast. However, mobile dunes with abundant *Medicago marina* still have *Ammophila arenaria*.
- **Dune grassland and scrub**. The more open dune grassland is rich in species. In Greece, for example, a dense grassland dominated by *Ephedra distachya* and *Silene conica* ssp. *subconica* is present. Dune heath can be identified by evergreen sclerophyllous shrubs or small trees such as *Juniperus phoenicea*, *Myrtus communis*, *Pistacia lentiscus*, *Spartium junceum*, *Arbutus unedo*, *Erica arborea* and *Quercus coccifera*. Here the vegetation is more similar to the maquis and gradations between dune and other forms of inland vegetation are not always easy to differentiate.
- **Dune slack**. These areas again support a different type of vegetation, which in places such as Bulgaria has many similarities with sites in northern Europe, for example *Phragmites australis*, *Schoenus nigricans* and *Agrostis stolonifera* are common. At some sites in the south, the wetter dune sand is colonised by pine woodland, which stands out in contrast to the dry, sparse vegetation found elsewhere on the dune ridges.
- **Woodland**. Natural woodland, where it occurs, is often dominated by *Pinus halepensis*.

A more detailed description of the form and vegetation of dunes throughout Europe is given by van der Maarel (1993) and Doody (1991).

7.3.2 Rare and alien plants

Dunes are a highly specialised habitat, especially in the early stages of development. There are a number of rare and declining species which occur there, some of which are restricted to this habitat. Amongst these, *Liparis loeselii,* a plant of dune slacks in South Wales, is declining (Jones & Etherington 1992), *Dactylorhiza incarnata* and *Parnassia palustris*, though relatively widely dispersed in the British Isles, are both rare species of dune slacks on the Wadden Islands and declining (Lammerts et al. 1995). In sandy areas and coastal dunes in Lithuania, *Centaurium littorale, Eryngium maritimum*, *Vicia lathyroides* and *Carex ligerica* are all at risk (Balevičienė at al. 1995). A total of 31 red data book species considered to rare, threatened or endemic have also been identified along the dune coast of Bulgaria (Petrova & Apostolova 1995).

Because of their inherent mobility sand dunes provide ideal opportunities for invasion by a wide range of alien species. Ranwell (1972a) reports the presence of 900 native and introduced species from a survey of 43 of the more important dunes in Great Britain. Many of these are associated with commercial forestry enterprises, such as the extensively planted *Pinus* spp., though others are derived from the weed flora of agricultural land. By contrast, only 400 species of native vascular plants were recorded by Salisbury (1952). Some of these exotics represent a threat to the native flora and fauna, though some may add to rather than detract from the value of the dune, as for example the attractive *Oenothera biennis* introduced to Europe from North America. Conversely, in the USA, alien species include *Artemisia stelleriana* (hoary ragwort), *Solidago sempervirens* and *Rosa rugosa*, the last also an invader of dunes in northern Europe, especially in Scandinavia.

7.3.3 Invertebrates

The variety of slope, aspect and vegetation structure, coupled with the dynamic nature of many systems, make dunes one of the most suitable habitats for a range of invertebrates requiring open conditions. They are particularly suited to warmth-loving species which bask on the dry warm slopes even in the more northern latitudes. Spiders show particular preferences to different zones, with open dune and dune heath having distinctive faunas (Duffey 1968). A study of the main invertebrate groups of sand dunes in Scotland (Welch 1989) showed that for Coleoptera (beetles) nearly half of the 88 species recorded were species of dry open habitats and 16 of these were found in coastal and dry sandy places. He concluded that many dune specialists require open dune habitats for their survival. Saltatoria (grass hoppers) and Formicidae (ants) also show clear preferences for different zonations within the dune structure (Handelmann 1998)

Hymenoptera (especially wild bees, digger wasps, spider wasps and other solitary wasps) build nests and are mainly sand-dwelling (Haeseler 1989, 1992). In a study of the group, 343 species representing 55.4% of a total of 536 were recorded on the north and east Friesian islands since 1950. Whilst the group declined on the north German plain; no decrease in species was observed on the islands which appear to be providing a refuge (Haeseler 1985). Brown & McLachlan (1990) make a distinction between dunes in arid areas, where the fauna may be especially limited and areas with greater moisture. The latter possess many more species, with invertebrates being particularly well represented.

In addition in open habitats the range of successions occurring over relatively short distances from open 'yellow' dune to 'grey' dune, dune slacks and scrub is important. The richest zones are those where clumps of stabilised vegetation occur in close proximity to areas of unstable open sand. These combinations allow species nesting in open soils to feed from the flowers of tall vegetation. Calcareous dunes are richer in invertebrates than acid ones (Kirby 1992.). The key to the survival of a full range of invertebrate species lies in the presence of a dune system large enough for all the stages of dune development to be present. It is also important for the more open conditions to be recreated as stabilisation takes place, thus ensuring the full range of conditions is maintained over the whole site.

7.3.4 Birds and other vertebrates

The dune fauna includes a wide variety of birds, e.g. gulls such as black-headed gull (*Larus ridibundus*) and especially in the north *Somateria mollissima* (eider duck) or the more widely dispersed shelduck nesting in the ground vegetation, and terns such as sandwich tern and little tern on open sands. Several birds of prey, such as harriers (*Circus aeruginosus*, *C. pygargus* and *C. cyaneus*) and owls (*Asio flammeus* and *A. otus*) may hunt and breed in larger dunes. The open cultivated and grazed machairs of the west of Scotland have high densities of nesting waders such as dunlin. Dune scrub is also an important habitat for a wide variety of passerines. At a few sites in the Mediterranean, tree-nesting species such as the squacco heron (*Ardeola ralloides*) and purple heron (*Ardea purpurea*) are present.

Rabbits are found in abundance in all areas and provide food for fox populations which occur in the mainland dunes. Polecat (*Mustela putorius*), weasel (*M. nivalis*), several species of bats, voles, mice and shrews occur in most of the areas, while roe deer (*Capreolus capreolus*) and red squirrel (*Sciurus vulgaris*) are less frequent.

Sand dunes are not known to be particularly rich in amphibians and reptiles. However, sand lizard (*Lacerta agilis*), natterjack toad, tortoises (*Testudo graceca* and *T. hermanni*) as well as edible frog (*Rana esculenta*) are amongst a number of increasingly rare and specialised species which occur in some.

7.4 Human activities and conservation

Historically dunes have been exploited in a variety of ways. Deforestation here, as in other terrestrial habitats, has significantly changed the nature of the vegetation. In Europe, areas such as the Łeba Bar in Poland, oak forest was present 2,000 years ago. Fire, probably started by human activity, destroyed much of this; replacement was by beech which expanded rapidly onto the mature forest floor (Piotrowska 1988). Today a natural climax vegetation with dune forests is rare. In northwest Europe high rainfall, combined with the long history of grazing by domestic stock or use as rabbit warrens, has usually resulted in stable dunes which are often dominated by grassland or heathland. In the drier Mediterranean, deforestation of oak/ash forest combined with grazing and burning has also changed the nature of the vegetation. Two of the more common communities, garigue (dwarf shrub communities) or maquis (bush communities), products of this management, are found on dunes.

Taken together, these activities show that dunes, particularly in northern Europe, have been subjected to considerable impact by human use. The changes that have occurred have not only modified the nature of their vegetation and changed the geomophological characteristics of many of the systems, but also massively reduced their surface area. Dune management today therefore takes place on sites which are both smaller and less dynamic than formerly. The discussion which follows attempts to unravel the complex interactions between the issues which influence dune development and describe how the human perception of dune management has changed in recent years. Some of these key issues are summarised in Table 7.18 below.

Table 7.18. Key issues for sand dunes, barrier islands, sandy spits and bars in Europe

HABITAT LOSS INCLUDING:	GRAZING MANAGEMENT:
Planting of forests of non-native trees;	Overgrazing/undergrazing.
Agricultural intensification, including reseeding and conversion to arable land;	TOURISM & RECREATION:
Housing development;	Car parks;
Building of airfields and use of defence lands;	Site of golf courses, including the building of greens, tees and fairways;
Building roads and other infrastructure including industrial development.	Trampling/erosion control;
ENGINEERING, SAND EXTRACTION & SEDIMENT SUPPLY:	Horse riding;
Sand extraction from the dunes and foreshore;	Lighting fires.
Offshore dredging from sand banks;	OTHER MANAGEMENT ISSUES:
Sea defence structures and coastal protection measures;	Invasive/alien species;
Remedial engineering including sand dune rehabilitation;	Burning, as a management tool;
Climate and sea level change;	Pipe laying, including gas and oil;
The importance of storms.	Marram gathering;
	Water tables and water abstraction;
	Saline intrusion;
	Acid deposition;
	The special place of the rabbit;
	Establishing grazing regimes.

7.4.1 Human activities

Historically, as has already been indicated, dunes have been exploited by humans in a variety of ways. They are particularly vulnerable to development including building houses, airfields and car parks, which cover the dune surface and interpose hard structures into a relatively soft and dynamic environment. Caravan sites and the development of golf courses, although less permanent, have also taken their toll. Agricultural developments, including conversion of dunes directly to intensive farmland and conversion of unimproved sandy pastures for arable use, though more easily reversed, also destroy dune grassland and heath. Sand extraction may reduce or destroy the area of dune altogether. The extent to which dunes have been affected by this wide variety of damaging and destructive human activities is summarised in Table 7.19 which is based on a survey undertaken by the European Union for Coastal Conservation.

Table 7.19. Estimated area of dune loss over the last 100 years in Europe Information compiled from a variety of European Union for Coastal Conservation sources (Doody 1995)

COUNTRY:	AREA (ha)	LOSS	CAUSES
Iceland	120,000	-	Erosion
Norway	>2,000	>40%	Development, reversion to scrub
Sweden	>2,000	-	Afforestation
Finland	1,300	-	Recreation, reversion to scrub
Poland	38,000	>80%	Afforestation, recreation
Denmark	80,000	38%	Afforestation (30,000 ha), recreation
Germany	<12,000	18-20%	Urbanisation, recreational development
Netherlands	48,000	32%	Afforestation (14%), tourist & urban development, water abstraction
Belgium	8,000	46%	Tourist & urban development
Great Britain	86,000	30-40%	Afforestation (14%), urban & recreational development, cultivation
Ireland	14,300	40-60%	Recreational development, erosion
France	280,000	40-78%*	Afforestation, tourist development, cultivation
Portugal	100,000	48-80%	Afforestation, cultivation, tourism
Spain	70,000	30-78%*	Afforestation, tourist & recreational development
Italy	<40,000	80%	Tourist development, urbanisation & afforestation
Yugoslavia			Tourist development
Albania	>2,000	No figures	
Greece	<20,000	40-80%	Tourist development, urbanisation
Romania			Afforestation, erosion
Bulgaria			Tourist development
Turkey	36,000	>30%	Afforestation, tourist development

*Loss on the Atlantic and Mediterranean coasts respectively.

Around 75% of the area of sand dune present a century ago is estimated to remain today. However, of this only 45% remains in more or less natural state. In the Mediterranean the figures are even worse with only 25% remaining intact. Over the last decades 30ha of dunes and beaches are thought to disappear every day (Delbaere 1999, EUCC Coastal Guide Web page http://www.eucc.nl/cstlguid/cstlguid.htm). Table 7.19 clearly illustrates the overwhelming effect of tourist development and afforestation on the status of Europe's sand dunes. The former is associated with the desire to build close to the beach to facilitate access to the sea and sand, the latter with the perceived need to stabilise eroding dunes. The following discussion takes each activity in turn and reviews their impact on the conservation of sand dunes and provides guidance on future management. Some of the key issues are depicted below (Figure 7.51).

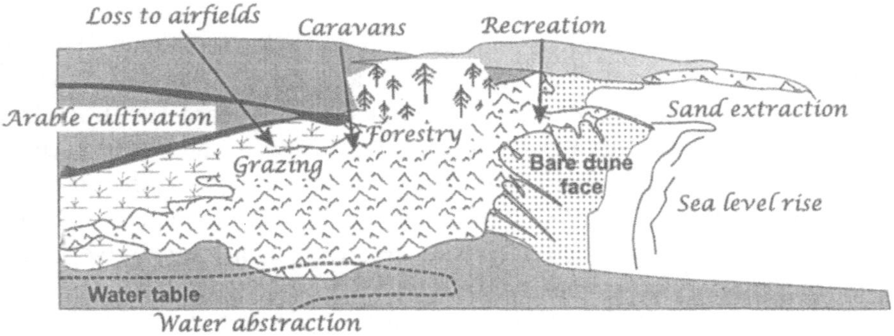

Figure 7.51. Key activities in the management of sand dunes

7.4.2 Tourism and recreation

The seaside holiday (the fore-runner of mass tourism) is said to have been invented by the English! It is attributed to a Dr. Richard Russell who came to Brighton in 1783 and published a book extolling the virtues of sea-bathing and drinking seawater. In 1783 Royal patronage encouraged growth in holiday usage. By 1822 when the famous Pavilion was complete, the resort of Brighton was already well established. The coming of the railway in 1841 stimulated further development and by 1900 a large proportion of the south coast was built up (Marsden 1947).

Sandy beaches and dunes have been especially favoured locations and 'lying in the sun' became an important part of the recreational activity. As Europe's population has become more affluent and travel cheaper, the upsurge in package holidays has resulted in the summer migration of the northern population to the warm south (Doody 1995). With sandy beaches as a prerequisite, the sand dunes often lying behind them, often provided locations for urbanisation and sand for building. As a consequence many dune landscapes have been severely degraded, especially in the Mediterranean. This has become particularly acute along both the Spanish and French coasts where tourist development has destroyed many dunes.

Recreational activity has other effects. The continued use of a sand dune for access has been the subject of a number of studies on the effects of trampling (Duffey 1967, Boorman & Fuller 1977). These studies show that the impact of trampling can have real effects, including initiating erosion and loss of species diversity. However, in all but a few cases the effects are relatively restricted, and other issues relating to dune stabilisation may be more important (see below).

7.4.3 Afforestation

Extensive afforestation has been used to stabilise dunes throughout Europe, affecting an estimated total of 25% dune loss (Tekke & Salman 1995). In Denmark planting began in earnest in 1883 and today 60% of State-owned dunes and 30% of private dunes are afforested (Skarregaard 1989). In Great Britain planting also began in the late 19th century and the process is illustrated by the case at Tentsmuir where the second biggest area of blown sand in Great Britain (after Culbin Sands, dealt with in Chapter 8, Figure 8.67) is almost completely covered with trees (including *Pinus sylvestris*, *P. contorta* and *P. nigra* var. *maritima*) (Figure 7.52).

In France maritime pines (*Pinus maritima*, *P. pinaster* and *P. pinea*) have been similarly planted over much of the dune landscape, particularly on the large expanses of dune which face the exposed Atlantic coast (Favennec 1998). In places these have been underplanted with oak, which on the acid sands of Les Landes includes *Quercus ilex*. Afforestation also affects many natural biotopes in Portugal. Historically, the planting of forests of *Pinus pinaster* (pinhals) has been a traditional way of controlling sand drift in dune areas. Today the planting of *Eucalyptus camaldulensis* has become very common and many areas including some dunes have been morphologically and ecologically destroyed as a result.

Extensive plantations in parts of the Mediterranean using the techniques employed by countries in the north and west did not take place on any scale until the 1940s (Fabbri 1997). In Turkey planting began on a small scale in 1888 and has continued from 1970 onwards on a massive scale. Between 1961 and 1990, it is estimated 10,672ha of coastal dune (29% of the Turkish coastal dunes) were affected. Natural vegetation is removed and replaced mostly by *Acacia cyanophylla* and *Eucalyptus camaldulensis* (both exotic), while *Pinus pinea* has been commonly used in recent years. (Uslu 1995).

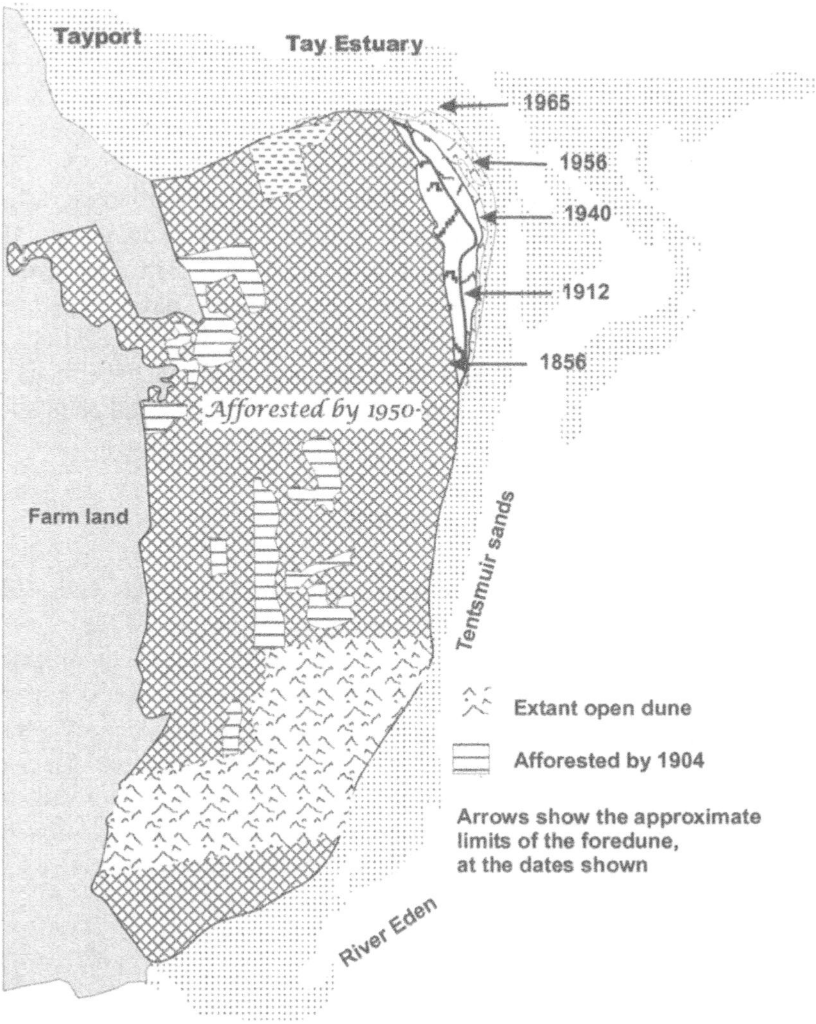

Figure 7.52. Progressive afforestation at Tentsmuir, Fife, Scotland

The planting of artificial forest composed of non-native pines not only destroys the natural dune vegetation as the canopy closes, but can result in an adverse change in the dune hydrology which may influence the vegetation at some distance from the forest. In Britain this appears to have resulted in the invasion of species-rich dune slacks by *Betula pendula* and *Pinus* spp. The overall effect is to destroy open sand dune communities, replacing them with a dense typically closed canopy with few of the native plants and animals surviving. Since mostly non-native trees are used, they seldom develop any significant nature conservation interest. There are exceptions and some of the pine woods in France are important for some larger animals such as wild boar, where larger scale and older plantings of well spaced trees occur. Whilst the Culbin forest, in Scotland, has pine martin (*Martes martes*) and red squirrel, the latter species also present in the pines on Sefton Coast dunes in northwest England. Both animals are rare in Great Britain. However the original dune habitat is lost and it is debatable whether these other interests compensate for the loss of the open dune landscape.

Whilst tree planting has helped to prevent sand blow affecting adjacent land and property; it has also 'fossilised' dune sediments. Today many dunes are 'over stable' showing a reversion of species rich forms to scrub and woodland (Doody 1989). In a few areas experiments have been tried to ascertain the efficacy of removing the woodland including the Sefton Coast, England (Sturgess 1992, Rooney & Houston 1998), in Denmark and Tentsmuir, Fife, Scotland, where a labour intensive programme coupled with goat grazing has been undertaken.

7.5 The importance of grazing

Grazing by domestic stock, rabbits and other herbivores has had a major influence on the vegetation of dunes throughout Europe. In the north and west it has helped to create stable or semi-stable dunes dominated by grassland or heathland. Changes in the nature of the vegetation occur where grazing pressure changes. As historical evidence shows, overgrazing combined with other human activities can lead to massive sand dune movement. Conversely the reduction or removal of grazing can result in the growth of coarse grasses and scrub. Both can have an adverse impact on species diversity and the presence of some of the more specialised dune plant and animal species.

7.5.1 Overgrazing

Overgrazing is a term which is usually applied when levels of domestic stock are kept on the land which are so high as to be detrimental to the conservation status of the habitat. When combined with burrowing by rabbits and rubbing of sand cliffs by sheep, destabilisation and deflation of large areas of dune grassland can occur. This is particularly true of the machairs of western Ireland, where paradoxically the erosion has exposed evidence of former settlements dating back several thousand years (Figure 7.53).

Figure 7.53. Deflation of 'machair' surface, Mannin Bay, Ireland. Note the humic layers possibly associated with prehistoric occupation

In the temperate northwest higher stocking levels are maintained through supplementary feeding or by providing animals with access to more productive land adjacent to the dune. This is particularly important during the winter, when feed is at a premium. From a conservation point of view artificially high stocking densities can have a number of impacts:

1. The greater use of the dune leads to trampling pressure which causes disruption of the surface vegetation and erosion. In the short term this may result in loss of mature grassland and heath and reversion of the dune to earlier stages of succession. If the grazing pressure is reduced or removed recolonisation takes place and new grassland or heath can become established;
2. Increased dunging provides an additional source of manure which can result in eutrophication of the dune surface and a change in vegetation type and structure.
3. The composition of the original feed may also have an adverse impact by introducing alien or more aggressive species not normally present as major constituents of the sward.

In most cases where grazing by domestic stock takes place, erosion poses a potential threat to the vegetation. This may be especially acute when the effects of high stocking densities are exacerbated by burning, as for example on the shores of southeast Turkey. Here the number of animals kept on the dunes is far in excess of their carrying capacity. As a result even in those areas which were still relatively free from other human interference, major loss of dune vegetation is not only detrimental to the dune but also to the animals which graze it. The solution which has been adopted includes the removal of the graziers and their stock. This is followed by planting to 'stabilise' the dunes using a variety of alien species such as the previously mentioned *Acacia cyanophylla* and *Eucalyptus camaldulensis*. Control of the grazing management regime would be a much better solution from a nature conservation point of view. This approach ultimately completely destroys the native dune vegetation as well as depriving the local population of their grazing lands.

7.5.2 Undergrazing

Undergrazing can be as detrimental to dune vegetation as overgrazing. When formerly grazed dunes are left unattended or stocking levels are lower than can remove the surplus vegetation, scrub and woodland can develop. In Norway for example, the change in management practice away from pastoral use has resulted in the growth of coarse vegetation and scrub. In Great Britain this has also led to a reduction in the conservation interest of a number of sites as the species-rich grassland is overgrown (Doody 1989). This effect reinforces the adverse impact of afforestation, referred to above and today many overstable dunes show loss of species-rich grassland and heath (a plagioclimax) to impoverished scrub and woodland communities.

The impact of this on the species composition of the Voorne dunes in Holland gives some indication of the nature of the change. Over a period of 25 years when grazing by domestic stock diminished, of the 579 plant species recorded between 1962-1970 the overall species number dropped to 540 in 1979. This included 104 losses and 65 gains, the losses being associated with open dunes, the gains scrub and other coarser species (van der Laan 1985). By contrast, experiments on reintroduction of grazing by cattle and ponies on the Meijendel dunes in Holland reversed the trend of decreasing species numbers (de Bonte et al. 1999)

7.5.3 The role of the r*abbit* (*Oryctolagus cuniculus*)

The rabbits was introduced to northern Europe by the Romans (probably from Spain) who brought it with them for food. The soft sand dunes of northwest Europe provided easy burrowing and in England, Holland and Denmark they were extensively cultivated. In Holland, for example, rabbits appear to have been introduced after 1280 (Wallage-Drees 1988). Here extensive sand dunes and "waredes" were leased to "Duinmeirers" similar to "warreners" in Great Britain where cultivation began early in the seventeenth century (Thompson & Worden 1956). In both cases the animals were used for food and fur. Individual colonies were enclosed and fed during the winter, an important activity to keep up the population levels.

Not only were rabbits important in the development of close-cropped grassland but also their burrowing activities helped initiate small-scale sand dune mobility. Thus for many decades the rabbit has been intimately bound up with the development of dune vegetation at many sites in northwest Europe. Ranwell (1972a) goes as far as to say "The structure of sand dune communities in Europe prior to myxomatosis was effectively the product of intensive rabbit-grazing.". In the early 1950s the species was effectively removed as a major component of the system as a result of myxomatosis. Since that time not only have rabbit numbers dropped drastically, and until quite recently remained so, but also there has been a major shift away from the pastoral use of the dune. During the period immediately after the decrease in rabbit grazing there appeared to have been an increase in plant diversity at some sites. Inspection of a number of sites suggested that species, suppressed by grazing, were able to flower more freely giving the visual impression of greater diversity (Doody 1989). By the 1980s questions were being asked by conservation managers about the loss of the open species-rich dune vegetation to scrub.

At Murlough Dunes in Northern Ireland, a rabbit warren from the 13th Century provides an illustration of the process. The rabbit population was decimated in the late 1950s. By 1967 when the National Trust took over the site the rabbit population had recovered but not before the establishment of extensive *Hippophae rhamnoides*. In 1976 a further decline in the rabbit population occurred and with it a further loss of open grassland and heath, as self-sown pine and other trees and shrubs invaded the site. Reintroduction of disease-free rabbits became a major, and successful, management tool from 1985 onwards (Whatmough 1995). An important consideration for rabbit numbers is the availability of winter feeding. At Murlough where the desire was to maintain numbers winter feeding took place. By contrast at Lindisfarne in northeast England, where overgrazing was a problem, fencing was used to exclude rabbits from their winter feeding areas which helped reduce numbers.

7.5.4 Setting grazing regimes

This discussion illustrates several points of importance to the dune manager. Firstly dune vegetation can respond very rapidly to changes in management. At Braunton Burrows it took less than 20 years from the time when it was thought there was too much mobile sand, to the point where major features of nature conservation importance were considered to be threatened by too little. Secondly, rabbits play a crucial role not only in grazing vegetation and preventing scrub development, but also in stimulating small-scale instability. Thirdly, without the presence of grazing domestic animals, and with a depleted rabbit population, dune vegetation quickly reaches a point where rabbits cannot easily re-invade.

Given the importance of grazing to the maintenance of sand dune grassland and heathland, establishing appropriate levels of grazing are critical. Generally for nature conservation purposes stocking densities on grassland of between 1.5 and 3.0 sheep per ha are adequate. On the Dutch Wadden island of Texel, dune heath is grazed at a rate a little over one sheep per ha (van Dijk 1992). Grazing by domestic animals is often the preferred management strategy on sites of high nature conservation value in temperate regions of Europe, whether as a continuation of existing practice, or the re-introduction of one which has ceased. The precise nature of the regime adopted will depend on an accurate knowledge of the existing pattern of vegetation, details of animal interest, and an indication of the history of development of the site and its recent management. These will help establish whether or not the present plant and animal communities are a true reflection of the site's full range of nature conservation interest. Where retrogressive successional development is identified it should also help determine the most appropriate form of management.

In addition to reviewing past regimes and where appropriate taking note of the approach adopted for other habitats, notably sea cliff grassland and saltmarsh, there are important practical points to bear in mind:

- *Do things slowly (to allow plants and animals to adjust);*
- *Don't provide supplementary feed (this may increase the nutrient levels);*
- *Do monitor and be prepared to adjust grazing levels.*

In some parts of the world grazing is not recommended. Based on their experience in South Africa and a review of sand dunes world-wide Brown & McLachlan (1990) go as far as to say "In some parts of the world dunes are actually used for grazing, a practice which should be abolished as quickly as possible." Given the significance of grazing to the maintenance of dune grassland and heath in temperate regions of northwest Europe it is doubtful if many dune managers would concur with this assessment. It also indicates an overwhelming concern for the prevention of erosion, which may not always be in the best long term interests of nature conservation on dunes as is discussed below (Section 7.9.4).

7.5.5 Dune restoration

The techniques used for restoring eroding dune landscapes for conservation purposes, when natural regeneration does not take place effectively, are well known and practised throughout the world (Figure 7.54). A practical guide to the variety of sand dune rehabilitation techniques such as sand-fencing and the like is given in Ranwell & Boar (1986). Burning of scrub in Denmark was also successful, but results suggest great care must be taken as intense burning can delay the re-establishment of dry dune heath (Vestergaard & Alstrup 1996). Various types of sand fences combined with planting of native grasses, notably *Ammophila arenaria* in Europe, are often used. Other more substantial structures may be used on beaches but can have detrimental effects, sometimes reducing rather than encouraging sand deposition. A key factor in devising any sand dune rehabilitation lies in understanding why the sand has been lost in the first place (Bird 1996).

Figure 7.54. Sand fencing on the New Jersey shore, USA

Coastal sand dunes provide simple and easy locations when making landfall for a variety of marine activities (e.g. oil and gas). In the UK the planning process can set conditions on proposals involving artificial disturbance that include reinstatement of the original dune vegetation. Such conditions resulted in detailed investigations being undertaken in northeast Scotland (St Fergus) where a series of gas pipeline landfalls were made. The results from these studies give an indication of successful techniques for restoration and the impact of the dune flora, fauna and physical attributes. Re-instatement of dunes along the pipeline corridor was most successful where *Ammophila arenaria* was planted into sand derived from the site itself, rather than into imported soils. It was also found that using naturally occurring grass species was preferable to the normally available agricultural seed mixtures. Factors outside the control of the developers sometimes had a major influence e.g. climatic effects and the presence of rabbits. Success was also impaired by local use which disturbed the rehabilitation areas (Ritchie & Kingham 1997). However evidence from this study also suggests that whilst rehabilitation techniques can be successful, reprofiling the dune and leaving it to its own devices may be the best option if it can be kept free from other human interference. In areas where dune sand deposition is inadequate to promote vigour in *Ammophila*, the use of rhizomes appeared to introduce all the elements needed to develop this "sand-stabilising" vegetation (van der Laan et al. 1997).

7.5.6 *Hippophae rhamnoides* (sea buckthorn)

Sea buckthorn is a native species on dunes in many parts of Europe. Because of its ability to stabilise mobile dunes it has been introduced at a number of sites where it is not thought to have existed before. In some of these areas it has encroached onto open dune vegetation to the detriment of the nature conservation interest. In Great Britain its impact on dunes was so great that a major review of its status and conservation implications was undertaken (Ranwell 1972b). Similarly in Belgium a rapid expansion took place in the late 1970s and 1980s at the expense of dune slacks which, without management, would have disappeared (de Raeve 1989). In Ireland its invasion of a large number of sites was thought to pose a serious threat to the native flora and fauna (Binggeli at al. 1992).

Methods of control use a combination of physical removal and treatment with an approved herbicide (Marrs 1985). This was successfully achieved at Braunton Burrows in southwest England where its removal was considered essential for the maintenance of the rich open dune vegetation on a site of national importance for nature conservation (Venner 1977). A combination of cutting by hand and machine, followed by burning of the cut material and spraying was used. A summary of the method is given by Ranwell & Boar (1986).

In Ireland total eradication was not considered appropriate as the mature stands provide shelter for a variety of birds and mammals, though measures were taken to control its spread outwards from these centres. In recognition of the threat posed by this and other invasive species of dunes, it was suggested that early removal of new colonies of *Hippophae* and other alien species should be undertaken before the problem becomes unmanageable (Binggeli et al. 1992).

7.5.7 Machair

Historical evidence suggests that the existence of the flat sandy plain, called machair, is directly attributable to human use. In the past the sand dunes showed considerable mobility as a result of management (or mismanagement) rather than natural climatic factors. Ritchie (1979) suggests that the extensive machairs of the Outer Hebrides (western Scotland) were "laid bare" by severe sand erosion and drifting as a result of overgrazing and intensive cultivation since the late 1600s. It is only in this century that traditional farming practices have helped provide an element of stability. Traditional cultivation is still extensively practised in the Outer Hebrides (Figure 7.55). [In Ireland the same formation is largely used as grazing land.]

Figure 7.55. Machair on South Uist, Outer Hebrides, western Scotland (Howbeg). Note the strip of cultivated soil in the mid fore-ground

The relatively low intensity of the cultivation techniques, including restricted use of herbicides and artificial fertilisers, allows expression of an older and richer flora, especially composed of arable weeds long-since eliminated from farmland elsewhere. It also provides suitable breeding sites (in the open cultivated sandy soils) for large populations of wading birds, principally dunlin and ringed plover. Densities can reach 200 pairs per ha in the most favourable habitats, though 100 per ha is more usual. These densities are far greater for this group of birds than for other habitats where they nest in other parts of Europe (Fuller et al. 1986). Nine species are present on the machair of Ireland though at a much lower density than on the Western Isles possibly because of the smaller size of the machair there and the absence of cultivation (Merne 1991).

Continued survival of this rich and varied wildlife heritage is intimately bound up with continuing traditional management of the land. This is very much influenced by the European Common Agricultural Policy, agreed between member states of the European Union. Increased subsidies for sheep can result in over-stocking of machair soils, leading to sometimes severe erosion as can be seen at a number of site in western Scotland and Ireland where deflation of the whole soil surface has taken place. This has led to a perception that erosion is a serious and wide-spread problem. From an historical perspective it is argued that although locally erosion is acute, the machair in Scotland is more stable than it has ever been (Angus & Elliot 1992). Though remedial action may be required locally other issues are also important. Amongst these tourist use including the destabilising impact of off-road vehicles, the removal of sand from the beaches (a problem widespread in Northern Ireland), dumping of old cars (sometimes used in a vain attempt to prevent erosion) and the invasion of *Petasites hybridus* (butterbur) on the island of Lewis all need to be considered.

7.6 Dune hydrology

Water relations in sand dunes are of special interest to the conservation manager. In addition to withstanding sand mobility, several dune-forming plants are specially adapted to deficiencies in nutrients and a lack of water. At the same time species-rich dune slacks are dependent on the level of ground water in the dune. The way in which dune water regimes affect vegetation development in dunes is dealt with by Willis et al. (1959), Ranwell (1972a) and Packham & Willis (1997, pp. 169-177). The following discussion looks at the implications for the conservation of dune slacks and the effects of drinking water production and precipitation on sand dune vegetation.

7.6.1 Dune slacks

Dune slacks commonly arise by the sand being blown away down to the level where it is stabilised by water (wet sand is not moved by wind). Consequently the water level of dune slacks is normally close enough to the surface for them to become flooded during wet winter periods. A lowering of the water levels can lead to the slacks becoming dry and also their level can rise by sand accretion; they may then be progressively invaded by woody plants such as *Salix repens* and eventually scrub and woodland. In sequence this seems to affect specialist species such as the rare orchid *Liparis loeselii*, which is an early coloniser of open dune slacks. It can be naturally out-competed as succession, including an increase in the depth of moss, takes place (Jones & Etherington 1992).

Grazing can be successful in reversing the trends, as is mowing. Sod removal is effective and in Holland mechanical diggers have helped provide a substratum closer to the dune water table from which recolonisation of dune slacks takes place (Jungerius et al. 1995). In Denmark evidence suggests that burning as a means of removing unwanted scrub is less intrusive and can be effective in areas where wet heath occurs since this prevents the superficial soils and seed bank from being damaged (Vestergaard & Alstrup 1996).

7.6.2 Drinking water production

The Dutch coastal dunes have been used to supply drinking water since 1880. The increase in abstraction resulted in a lowering of the water table and by 1900 wells had to be sunk 3-4 metres down. Further increases in demand caused shrinkage in the volume of freshwater and brackish sea water began to infiltrate the abstraction wells (van Dijk 1989). Infiltration works have been constructed to allow water to be pumped from the Rivers Rhine and Meuse to recharge the aquifer (Huisman & Oltshoorn 1983). In addition to the direct loss of habitat to building associated with the water-works the initial lowering of the water table caused the dune slacks to dry out with a loss of some 80 species including rarities such as *Parnassia palustris* and *Epipactis palustris* and an increase in more invasive species such as *Hippophae rhamnoides*. Following infiltration the original plants did not return. This was due to a number of factors including: the height of the water table which did not revert to its original level (it was usually higher); water level fluctuations were more rapid and not linked to seasonal factors; the ground water flow was accelerated and there was an increase in the nutrient status of the ground water (van Dijk 1989).

7.6.3 Nitrification

Acid deposition and pollutants have been implicated in changes in the vegetation of sand dunes. Studies on the dune system on the former island of Voorne in the Netherlands, which were made over a period of 25 years suggested that the disappearance of *Eryngium maritimum* and *Euphorbia paralias*, both dependent on sand deposition, and the increase in *Hippophae rhamnoides* scrub was attributed to acid deposition and pollution (van der Laan 1985).

7.7 Golf courses

Golf courses are a feature of many sand dunes in the United Kingdom and in Ireland and they are also increasing throughout Europe. The extent of loss of the dune flora and fauna varies on different courses but habitat fragmentation through the destruction of the semi-natural vegetation occurs when greens, tees and fairways are established. These artificial dune habitats are much less rich in species. In a study in Jersey semi-natural dune vegetation had 30 to 40 species per m^2 compared with 6 to 10 species per m^2 on fairways which have received fertiliser and herbicide applications as well as irrigation (Ranwell 1975). All golf courses contain areas of 'rough' and in some cases these are also intensively managed which restricts the diversity of the vegetation. By contrast, on a few sites relatively large and intact areas of 'rough' remain supporting important plant and animal communities. In these areas management intensity is low, perhaps involving mowing on an annual cycle or grazing at low levels.

As the popularity of golf has grown so have the number of courses. In the Netherlands for example, there were 61 courses in 1989, of which 6 were situated on sand dunes. There were plans for 67 more with 6 of these on or adjacent to sand dunes (van der Zande 1995). New tourist urbanisation often includes a golf course as an essential part of the attraction for a 'high quality' resort. The dunes of Maspalomas in Gran Canaria have suffered destruction from extensive tourist urbanisation. An extensive golf course shows up as a green oasis contrasting with the damp slacks where *Phoenix canariensis* survives. No less than 27 new courses have been constructed in the Algarve, Portugal. The development of the greens, tees and fairways, coupled with the requirement for water, further degrade the surrounding natural environment.

New courses will always destroy part of the dune landscape vegetation, particularly when the faster more manicured courses favoured by American style golf, are promoted. However this must be tempered by the fact that in many areas more damaging developments might well have taken place. At the same time there are opportunities for more sympathetic management. Today a number of existing courses, especially some of the older links courses in the UK, still retain important sand dune communities. These include rarities such as *Himantoglossum hircinum* (Figure 7.56) which thrives on the Royal St Georges golf course. It is specially protected when major tournaments take place, especially for The Open Championship, which is played there from time to time. As Fred Pearce put it in the *New Scientist* in 1993, "...before bulldozing starts (on a sand dune), perhaps the course designers should...have a look at Royal St Georges".

Figure 7.56. Lizard orchid (*Himantoglossum hircinum*), Royal St Georges golf course, Kent

7.8 Sand dunes, climate change and sea level rise

Sandy beaches are perceived to be eroding along a substantial proportion of the World's coastline. This has been estimated to be 70% over the last one hundred years (Bird 1985). The reasons for this situation are complex and cannot be attributed to a single cause. Reduction of available sediment for new dune formation is a key factor. Marine aggregate extraction, protection of eroding cliffs, canalisation and damming of rivers as well as relative sea level rise and the availability of reworked Holocene sediments, may all play their part.

In Denmark the last major movement of sand occurred between 1480 and 1780. Whilst this is attributed to over-use by the inhabitants of the dune, climate change during the 'little ice-age' may help to explain the presence of the large amounts of sand which would have been required to cover the dune. It is suggested that sea level change may have been as important, mobilising sediments offshore and making them available for dune building

(Christiansen et al. 1990). In Holland three stages in dune development appear to correlate with climate history, storm surge frequency and coastal erosion brought about by a rising sea level (Klijn 1990).

Because of their high degree of mobility if left to themselves, the foreshore and fore-dune move in response, as sea level rises or falls (Bird 1993). Where relative sea level is rising the effect is likely to be beach erosion and migration of the dune front landward (Bird 1996). These effects may be even more pronounced when the climate is colder and drier and human actions help to create instability in the body of the dune. This has important consequences for habitation on or adjacent to eroding beaches as they may be overwhelmed during storm periods. Where there is a restricted sediment supply and an artificial landward boundary has been imposed by building, storms can be particularly damaging, as happened to the beaches of Ille-et-Vilaine, northern Brittany in 1990 (Regnauld & Kuzucuoglu 1992).

7.8.1 The sand dune 'squeeze'

The Polish coastline provides a classic example of the sand dune 'squeeze'. Planting of the dune landscape began at the beginning of the 19th century following destruction of the original oak forest about 2,000 years ago. The erosion, which the planting of pine forests was set to overcome, reduced the open dune habitats such that they now represent a much impoverished nature conservation resource. Despite the massive and long term commitment to stabilisation the dunes continue to erode (Figure 7.57), suggesting that the natural erosive forces may be of overriding importance (Piotrowska 1989).

7.8.2 Sand dune management - the nature of mobility

The combination of direct loss of sand dune to the causes identified above, coupled with loss of sandy beaches, results in the habitat being squeezed into an ever narrower zone. Because of the highly dynamic nature of the habitat itself (at least in the early stages of development) and the fact that so much attention has been paid to ensuring dune stability, dune conservation management today takes place in many areas against a background of a reduced and decreasing area of open dune landscape.

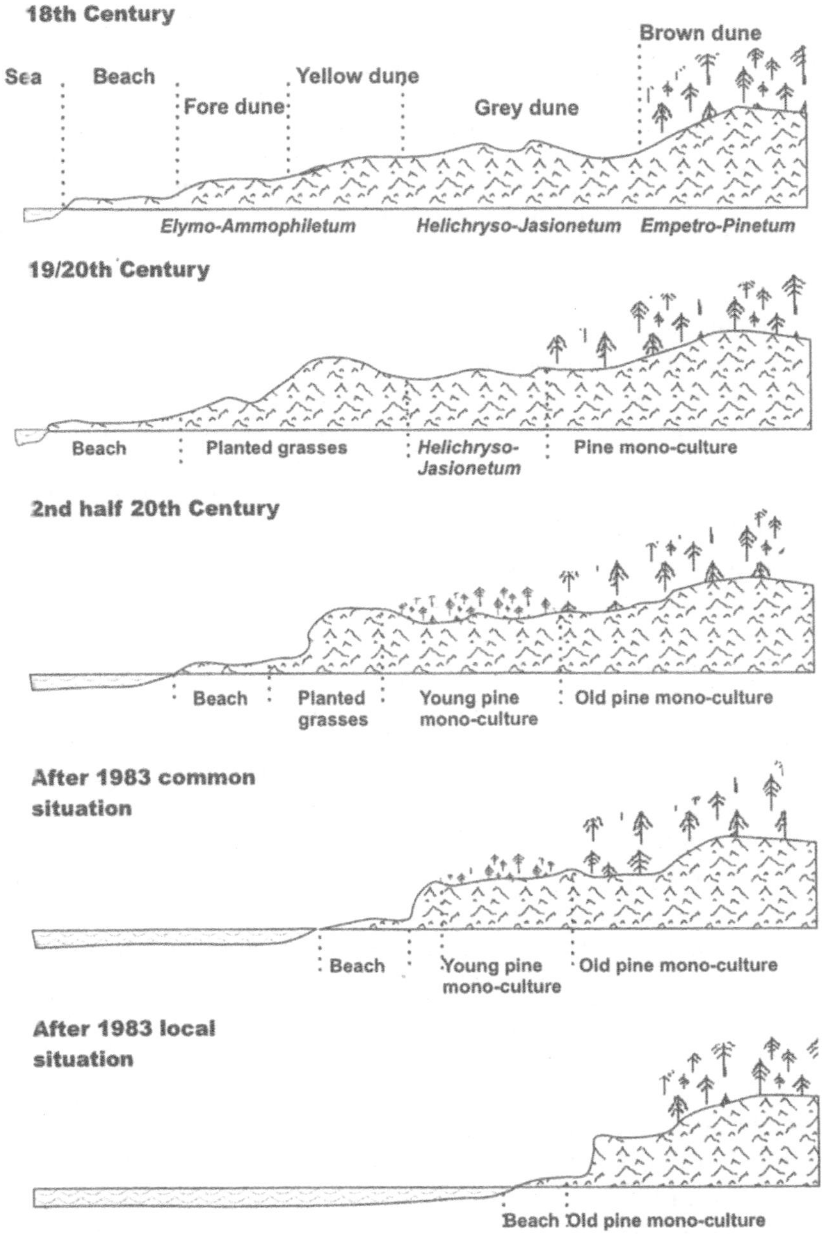

Figure 7.57. The sand dune 'squeeze' in Poland (after Piotrowska 1989)

On the east Friesian Island of Spiekeroog, for example, management since 1629 has seemingly been directed towards producing 'stability' in the dune (Table 7.20)

Table 7.20. Dune stabilisation periods on the island of Spiekeroog (Isermann & Cordes 1992)

Date	Activity	Date	Activity
1629	Prohibition of *Ammophila* cutting	1860	Afforestation
1711	*Ammophila* planting	1880	Extermination of rabbits
1828-1910	Increased tourist use; a shift from agriculture, lower cattle numbers	1892	Introduction of *Hippophae*

Disentangling anthropogenic effects from more natural forces is essential when attempting to predict impacts resulting from global warming or other destabilising effects. Braunton Burrows in Devon provides an important illustration of how perceptions of the problems of dune conservation have changed. During the Second World War part of the site was used as a military training area, in particular as a practice ground for the Normandy landings. As a consequence large areas of dune vegetation were laid bare and the sand became very mobile. In the early 1960s the Ministry of Defence handed the site back to the owners. The combined effects of military use and uncontrolled rabbit grazing and burrowing, had resulted in the creation of hundreds of acres of bare sand.

In recognition of the danger of major loss of sand to the system, between 1960 and 1963 a large scale programme of marram planting was undertaken. During this period *Hippophae rhamnoides* also colonised the dune system (following an earlier failed attempt to introduce it in 1933) speeding up the stabilisation programme. The outbreak of myxomatosis in 1953 was seen as a major aid to this work. The combined effects of the artificial stabilisation programme, the growth in some places of *Hippophae* and the loss of rabbits had the desired effect of stabilising the dune.

Part of the site became a National Nature Reserve in 1964 when nature conservation management became the responsibility of the Nature Conservancy. By the late 1960s it was recognised that the invasion of *Hippophae*, at the expense of the species-rich dune vegetation, had reached such a state that a major control programme needed to be undertaken over the whole site (Venner 1977). Despite the success of this control programme, the growth of coarse grasses and the continuing natural stabilisation of the dune system in the 1980s and 1990's (Packham & Willis in Press), caused further losses of the communities for which the reserve was established. This led to acceptance of the re-introduction of grazing animals as the only cost-effective way of restraining the growth of scrub and re-establishing the extensive rich dune vegetation.

There were encouraging signs that trials using Soay sheep at Braunton Burrows had, in the short term, helped recreate open dune grassland habitat. However grazing has not at present been introduced on a wider scale. In another part of the site a major blow-out is helping to create a new dune slack as a wandering dune moves over the system. No attempt is made to stabilise this blow-out has been made to date. In 1998 the re-introduction of cattle and sheep was initiated in the context of an agreed 10 year Braunton Burrows Management Plan (Packham & Willis in Press).

7.8.3 Catering for tourists - caring for dunes?

In Holland and Denmark the use of dunes for water catchment, sea defence and recreation has largely prevented extensive urbanisation (Jensen 1995, van de Zande 1995). However, recreational use brings its own problems and whilst the perception of the dune manager may sometimes overemphasise the impact of erosion, in densely populated areas and at particularly popular locations, recreational use must be controlled. In places where vehicles and large numbers of people can gain easy access to relatively small areas, more serious and irreversible destabilisation can occur. Limiting or controlling access is the solution here.

Tourists require a wide range of facilities which cater for the traditional beach holiday as well as more active pursuits such as walking, running and natural history. A key aspect of any visitor control programme lies in the location of car parks. The Meijendel Dunes in Holland are a highly popular recreational area near the Hague. Here the location and size of car parks is used to control visitor numbers, their access and hence overall impact on the dune habitat. The traditional beach holiday is catered for by providing a large car park at one end of a sand dune system with access to the beach. Access to the rest of the site is by foot only. As a consequence the pressure on the major part of the dune is low and only small scale remedial action is needed to control sand movement (Bakker 1998).

Many of the world's surviving sand dunes retain important wildlife interest. They may also be significant areas for recreation, sea defence, forestry and agriculture. Although these uses are not always compatible with the maintenance of the nature conservation interest, they provide an important additional justification for their protection and management. These less destructive uses can be accommodated without a permanent threat to the wildlife interest, if appropriate management decisions are taken. On the majority of the surviving dunes in Europe the development of an integrated strategy for dune conservation must recognise that human use, including recreational activity, should be considered as an integral part of the management process. This is entirely consistent with the political goal of 'sustainable development' which was expressed at the Rio Summit in 1992. When linked with the maintenance of biodiversity a strong case for dune conservation can be made.

7.8.4 Erosion and accretion - the case for mobility

Dune managers, including those with responsibilities for nature reserves, often perceive dune erosion as 'a bad thing'. Given the historical use and misuse of dunes described above, this conclusion is understandable and is reinforced by the increased pressure caused by the loss of habitat through afforestation, the building of houses, tourist urbanisation, including caravans and roads. The remaining areas are, as a consequence, considered to be even more precious so the protectionist instinct becomes more intense. It is perhaps not surprising that so much management activity and expertise have been concerned with erosion control.

Problems of erosion are real and require solutions. However, the techniques are well established, both in relation to rehabilitation of the dunes themselves and the management of people. Except in a relatively few cases, particularly in the small exposed sand dune sites or adjacent to car parks, caravan sites and other access points, erosion is localised and traditional management techniques can be applied. Even in the Dutch dunes where very restrictive controls have been followed for many years, studies of the effects of mobilisation of inland dunes showed that blow-outs rarely exceeded 30m in length before natural stabilising factors such as vegetation succession and soil maturation (aided by the growth of algae on the surface of the sand) came into play (van der Meulen & van der Maarel 1989).

Cycles of erosion followed by periods of stability are all part of the natural development of dunes and essential to the maintenance of their wildlife value, including the exceptional invertebrate fauna. Although human use (or misuse) can be a major destabilising force, once the human factor is reduced or removed more stable dune habitat is quickly formed. Periods of destabilisation followed by stabilisation have characterised dunes in many areas throughout history. The origin of many dunes, including the 'young dunes' of north and south Holland (between 1200 and 1600), is attributed to a combination of deforestation and periods of heavy storms (Jelgersma et al. 1970). In Denmark major sand movement involving the large scale displacement of houses, farms and villages occurred in the 16th and 17th centuries (Skarregaard 1989). Throughout Europe the principal agents of sand instability have been burning, over-grazing by domestic stock and the removal of vegetation for animal bedding and thatch. Merino et al. (1990) describes the "critical" role of human intensification of land-use around 1628 in the "desertification" of coastal sand in southwest Spain.

From this perspective it would appear that most of the present dune landscapes derive partly from previous human use (and over-use). This use may have helped create the rich open mosaics so prized today. Thus far from being a problem, new cycles of sand movement, including those caused by excessive trampling or other destabilising force, if followed by a period of freedom from human use, may help recreate degraded or overgrown vegetation. The key to developing such a management policy lies in understanding the interrelated nature of the system. Rapid change is the norm and, far from being fragile, sand dunes erode and accrete in response to storms, sea level changes and the like and in this sense are quite robust. Thus dune dynamics can, and should, play a part in their management for nature conservation.

A decision to open a stabilised dune surface deliberately to the action of wind and rain may be difficult. Other legitimate interests such as foresters or coastal engineers may see their roles compromised. It could be equally difficult to persuade some elements of the conservation movement that the work should be undertaken. This point is illustrated at Ainsdale National Nature Reserve on the Sefton Coast. Here over-stabilisation of the habitat and encroachment of scrub and woodland led to a decision by English Nature (the statutory organisation responsible for management) to remove some of the pine forest. This was felt to be an important stage in securing the long term survival of the rich dune and dune slack habitat, together with their notable animals the sand lizard and natterjack toad. Local views are polarised between those who recognise the value of the dune in its dynamic form (Houston 1992) and those who see the woodland as a habitat for the rare red squirrel. In this case the survival of the red squirrel has become a major focus for local opposition to the tree felling programme.

Some form of destabilisation will be beneficial to those dunes where grazing (including that by rabbits) is curtailed. The problem of the sand dune manager may well lie in his perception of the size of any erosion features caused by human activity. When viewed at or near the point of sand movement, 'the blow-out' may seem to be a major threat. However, looking from a wider perspective, as may be obtained from aerial photographs or satellite images, it may only affect a limited area in proportion to the system as a whole. Acceptance that this sand movement forms part of the 'natural' process, with positive benefits for nature conservation, may require a fundamental change in attitude.

Left to their own devices and in an environment where there are changes in sediment availability, sea level and climate, sand dunes respond by adapting to the new environmental conditions. Not only is 'today's blow-out tomorrow's dune slack' but in its turn it can provide suitable conditions for the survival of rare species such as the natterjack toad or *Liparis loeselii*.

Many habitats may be short-lived but so long as the dune is large enough and continually responding to the driving force of dynamic processes new areas will be created and a continual cycle of change will ensue. This is entirely contrary to the present situation on so many dune systems where human intervention has imposed artificial constraints. In the future conservation management should aim to provide larger and more dynamic dune fields where interference is minimal.

Constraints, such as golf courses, may make such losses unthinkable. However, a more dynamic approach could be more sustainable in the long term than seeking to protect the tee where a famous golfer shot a first 'hole in one'. Better still, the 'fixed' dune landscapes, such as the massive artificial forests which abound in Europe, might be considered suitable targets for restoration by removing the trees. In some parts of Europe mobile dunes are a major tourist attraction, as on the Łeba dunes in Poland. Fire and forest clearance initiated their dynamic condition. This is maintained by the numerous visitors who wander up and down the dune crest, which moves at up to 10m per year (Rotnicki 1994, Figure 7.58). Perhaps this could be a paradigm for the sand dune manager.

Figure 7.58. Massive mobile dune in the Slowinski National Park, Poland - the Łeba Bar

8. SHINGLE BEACHES AND STRUCTURES

8.1 Introduction and scope

Shingle beaches are widely distributed around the world. They are composed of deposits of pebbles lying at or above mean high water. These most often occur as fringing beaches which are subject to periodic displacement or overtopping by high tides and storms. Waves determine the position of the sediment on the beach. Deposits may be reworked in front of the shore or moved in parallel to it by longshore drift, before being thrown up onto the beach by storm waves. Small foreshore ridges are deposited at the limit of high tide and form the fringing beaches which may occur on their own or as precursors to more permanent ridges.

Stony banks are much less frequent, but where they occur they can become significant features in their own right. Sequences of ridges can be piled up against each other in parallel lines reflecting the prevailing direction of storms. The material may be derived from cliff erosion, transported from rivers draining highland areas, or from offshore deposits of glacial material which may include erratics.

Stability and availability of water are overriding determinants of the vegetation. Only a few specialised species are able to colonise the early stages of succession. Several of these are ephemeral e.g. *Cakile maritima* and *Atriplex glabriuscula*. Specialised plants such as the northern *Mertensia maritima* and the more southern *Lathyrus japonicus* and *Crambe maritima* (Figure 8.59) are much rarer. As the ridges become stabilised, vegetation develops as fine material accumulates between the pebbles. This provides more favourable conditions for plant growth and a range of communities occur including grassland, heathland and scrub.

Figure 8.59. Crambe maritima; a long root system enables the plant to find water and withstand disturbance

The intensity and type of human use vary widely. In many northern areas important and virtually untouched beaches and structures occur. These include narrow beaches which front sea cliffs and fringe sea lochs (fjords) or occur as wide expanses of parallel ridges in exposed locations. Human influence is not entirely absent even in unpopulated areas. In Alaska, for example, the large and seemingly natural shingle structure at Cape Krusenstern shows evidence of cultures living in association with it over many hundreds of years (Giddings 1977). More usually permanent destruction takes place as large areas of shingle often provide accessible resources of gravel. In this latter situation the surface of the shingle can be obliterated and the whole structure destroyed. In contrast, artificial barriers can, on accession, arrest so much shingle that a vegetated beach is formed. This occurred after 1972 at Black Rock beach, Brighton, East Sussex, when the western harbour arm of the marina was constructed. Within a short period a new beach developed and to date 54 species have been recorded, but a number have been lost in storms (Packham & Spiers in Press).

This chapter deals with the general issues associated with the conservation of shingle shores and structures. A more detailed review of the communities and their development is provided by Packham & Willis (1997, Chapter 8). Because of the relative rarity of major structures, much of the discussion which follows centres on a small number of individual sites. In addition to considering management specifically for nature conservation this chapter also looks at some wider issues such as sea defence.

8.1.1 Habitat definition

Shingle shorelines and structures develop in areas of high wave energy. Their precise form is dependent on the wave direction, the incidence of storms and the availability of sediments. Shingle shorelines occur as fringing beaches where storm waves throw pebbles on the beach which are subject to periodic displacement or overtopping by high tides and storms. On some shorelines varying amounts of sand are interspersed in the pebble matrix and gradations to sandy shores may occur.

Beaches are described as being "**accumulations of sediment deposited by waves and currents in the shore zone**" (Bird 1984). Shingle beaches and structures are defined by the size range of the particles and are usually composed of **deposits of pebbles ranging in size from 2mm - 200mm** in diameter (Randall 1977).

Shingle structures are accumulations with a sequences of ridges which become piled up against each other in parallel lines. The predominant particle size is the same as for shingle beaches. The difference relates to the frequency of displacement by storms. They range in size from small features with saline 'percolation ponds' to much larger structures where the older ridges become completely separated from the sea by new deposits thrown up during storms or on shorelines where relative sea level is falling.

8.1.2 Habitat type

There are five main types of shingle which provide a substrate for permanent shingle vegetation. The first three are regularly washed by sea water spray and storm waves and vegetation is restricted. These include fringing beaches, shingle spits and bar/barriers. The more complex structures include cuspate forelands and offshore barrier islands. These may occur as extensive areas of parallel ridges such as Cape Krusenstern, referred to above, or Dungeness in Kent (both cuspate forelands). The shingle structures of Scolt Head Island in Norfolk (a barrier island) and the Culbin shingle bar in Moray occur in conjunction with sand dunes and saltmarshes to form extremely important geomorphologically complex ecosystems. Figure 8.60 below shows the main type of formations derived from examples in Great Britain.

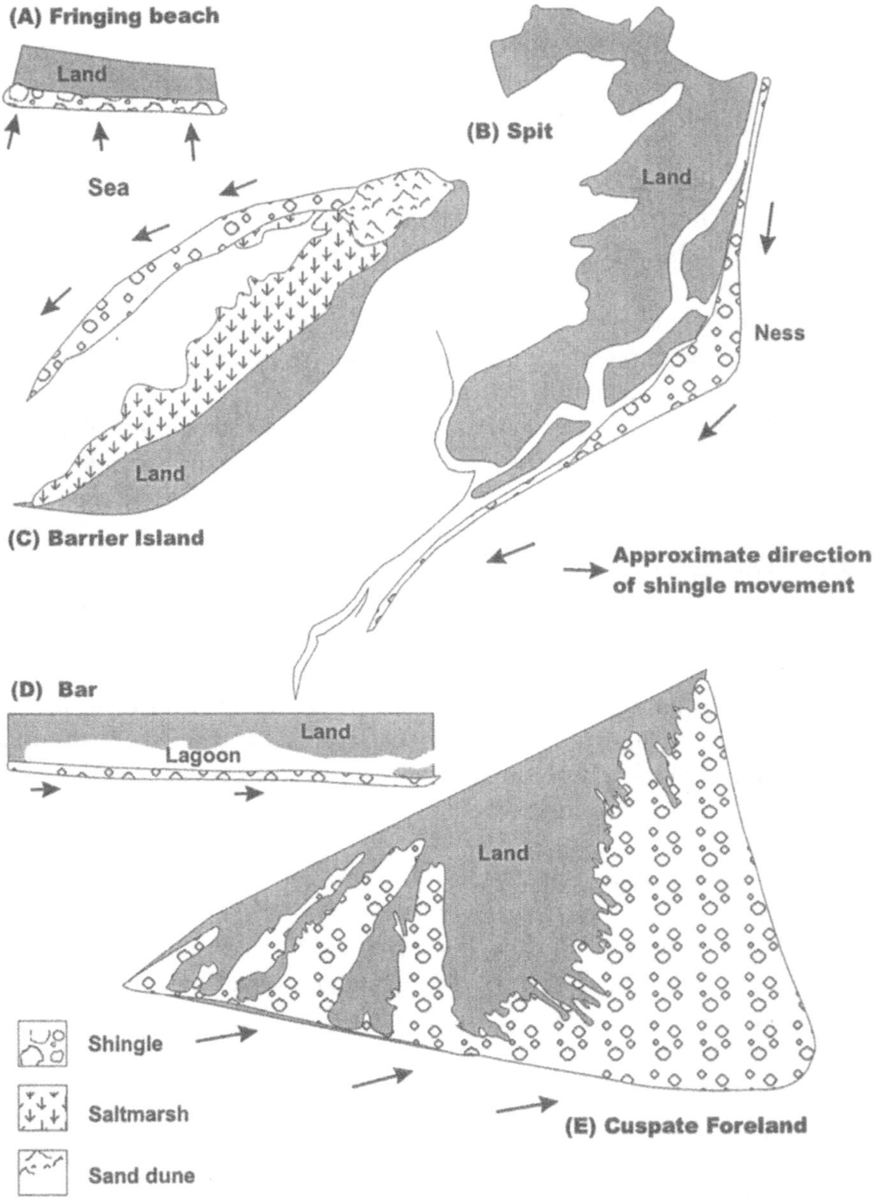

Figure 8.60. Shingle structures based on examples from Great Britain. Not to scale

- **Fringing beaches**. The most common form, usually occurring as a simple strand of shingle in contact with the land. They are mobile and vegetation is usually sparse but very characteristic. Fringing beaches occur all around Britain and in many parts of Europe, especially in Scandinavia.

- **Shingle spits**. These occur where coasts have an irregular outline, especially where there is an abrupt change in direction. Recurved hooks form when the shingle is deflected by waves around the end of the spit. Major examples are rare. Orfordness, Suffolk, is probably the best developed though examples also occur at Sillon de Talbert, Brittany.
- **Bar/barrier**. Similar in form to a spit, but crosses an estuary mouth or bay and ultimately encloses a freshwater or brackish lagoon (Slapton Beach, Devon).
- **Cuspate foreland**. Large structures formed from a series of parallel ridges. These occur when shingle is piled against an existing spit or fringing beach as at Dungeness in Kent, Korshage in Denmark, and Rügen, Germany.

Offshore barrier islands are shingle deposits in shallow water, often linked to other coastal habitats. They can originate as spits detached from the mainland as in the Culbin Bar, Scotland, North Norfolk, England; and the Estonian Islands. Special and complex structures can develop involving both bars and spits, as in the sea lochs around the Shetland islands (Figure 8.61).

Figure 8.61. 'The Houb', a complex double shingle structure enclosing a saline lagoon, Shetland Islands, Scotland

8.2 Habitat distribution

Shingle beaches are common and occur throughout the world where there is sufficient wave or tidal energy to transport the larger sized pebbles. Shingle structures of any size are rare and Europe is one of the few areas where they are known. The others are New Zealand (Single & Hemingsen in press), gravel barriers on the Atlantic coast of Canada (Orford et al. in press) and Japan (Nakanishi 1982, 1984). Shingle beaches and structures are also widely distributed around the coast of Great Britain (Sneddon & Randall 1993a, b, 1994a, b) and Ireland. Elsewhere in Europe little is known about the distribution of shingle structures though a brief overview has been prepared (Randall & Doody 1995, Figure 8.62).

8.2.1 Habitat distribution in Europe

Coarse shingle beaches occur in the high and low Arctic (Iceland, North Norway, Russia). In south and central Norway there are narrow sand and gravel beaches in moderately exposed inlets with drift and mollusc shell banks and some larger structures of shingle and boulders also occur at Tromøya, Jonfruland and Mølen. These include 'impressive' shingle formations which are derived from glacial deposits laid down some 10,000 years BP (Lundberg & Lasvik 1993). In north Sweden and Finland the freshwater of the north Baltic, the severe winter climate and land uplift have resulted in extensive stony shores, often virtually devoid of vegetation except in sheltered sites. A classic site occurs on the protruding eskers at the mouth of the Dalaven River, Billuden, Uppland. There are extensive 'flintstone fields' in Baltic Germany, for example on the island of Rügen. In Denmark there is a sandy gravel coast on the north of the Korshage peninsula, Sjaelland (Randall & Doody 1995).

Further south and west there are extensive beaches and a number of important structures occur in Great Britain. Ireland has a relatively large number of small scattered sites. In northwest France there are deposits of chalk-flint shingle near Boulogne, the "blue" pebbles at Cayeaux sur Mer south of the Somme, and south of the Seine estuary. Other locations include Sillon de Talbert in northern Brittany where a natural bank extends 3km into the sea (Géhu 1960). Generally shingle appears to be absent from much of the Mediterranean, though Corre (1993) reports coarse sand shores in Mediterranean France which may verge on shingle habitat. Lovric (1993a) describes minor beaches of coarse cobbles 3-12cm diameter on the coasts and islands of Croatia, e.g. Pag Island. There are also small examples in Albania and Greece. There are no references to the presence of shingle on the shores of the Black Sea though the source map from the European Union for Coastal Conservation shows a few locations on its eastern shore.

Figure 8.62. Distribution of shingle beaches and structures in Europe based on Randall & Doody (1995) with modifications

8.2.2 Habitat distribution in Great Britain

Shingle beaches and structures in Great Britain are relatively well known. Approximately a third of the coastline of Great Britain is fringed by a shingle or sand/shingle beach, much of it devoid of vegetation. During 1987 and 1990 more detailed surveys of the vegetation of all the larger shingle structures were undertaken (Sneddon & Randall 1993a, b, 1994a, b). These studies showed that although a considerable length of coast was fringed by shingle beaches, natural vegetated shingle structures are limited to only 7,200ha of stable or semi-stable vegetated shingle.

8.3 Nature conservation value

The nature conservation importance of shingle derives from its rarity, geomorphological interest, unique vegetation and associated fauna (especially invertebrates). Although fringing beaches are common in the north and have species which are ubiquitous, there are rare plants and invertebrate animals unique to the habitat, making them unusual and valuable places. However, it is the form of the structures and the sequence of vegetation which has developed in response to changing patterns of climate, sea level change and human occupation that are of greatest interest.

8.3.1 Successions and transitions

The earliest stages in shingle beach formation are characteristically devoid of vegetation because of the constant movement of the pebbles by wave action. The vegetation is composed of a high proportion of annual or short-lived perennial species and is very distinctive. The European Union Habitats Directive identifies two communities. The first; **Annual vegetation of drift lines** (Code 17.2) includes *Mertensia maritima* prevalent in the north and *Glaucium flavum* and *Matthiola sinuata* in the south. Mobility is an over-riding consideration and the species which colonise are able to withstand periodic disturbance. These include *Lathyrus japonicus* and *Crambe maritima* colonising the shingle just above the high water mark where the shore may remain stable for periods of 3-4 years (Scott 1973). *Crambe maritima* is special significance as it helps define three sub-types of the second major community **Perennial vegetation of stony banks** (Code 17.3) for the Baltic, Channel coast and the Atlantic respectively (European Commission 1999c).

Species are also tolerant of saltwater inundation as the beaches are often overtopped by waves breaking over them. As the shingle is pushed further out of the reach of all but the most violent storms a more complete vegetation cover develops. Above this zone at some sites where sea spray covers the more stable plant communities, there is a high frequency of salt tolerant plants such as *Armeria maritima* and *Silene uniflora*. These may exist in a matrix with an abundant lichen flora.

During the development of a shingle structure, smaller pebbles tend to be thrown to the top of the shore. Coarser sediments remain lower down the beach, as the waves carrying the material move landward and then recede. Where the structure has become sufficiently stable to allow fine material to accumulate and plants to grow a range of communities develop including grassland, heathland and scrub. Established ridges contain finer sediments than the intervening 'lows' and it is this pattern of alternating sediment which in part determines the eventual vegetation distribution.

In more sheltered situations terrestrial plant communities become established. Heath vegetation with *Calluna vulgaris* and/or *Empetrum nigrum* occurs on the more stable shingle structures, particularly in the north. On the largest and most stable structures the sequence of vegetation includes scrub, notably *Cytisus scoparius* and *Prunus spinosa* best represented at Dungeness in southeast England. A characteristic sequence can occur with vegetation developing on the tops of ridges but not in the hollows. Evidence from the UK suggests that there is a succession involving the growth and degeneration of shrubs such as *Cytisus scoparius* (Ferry et al. 1989). These communities, together with lichen-rich grassland and heath which occur in exposed locations subject to saltspray, harbour a rich invertebrate fauna including many specialist species.

An unusual form of this community occurs in East Anglia where the shingle exists in a matrix of sand. It represents an intermediate stage between shingle beach and a sandy fore-dune with characteristic species from both. Elsewhere a few shingle beaches are composed of marine shells. Little is known about these and they are not considered further here. A review of plant community development based on examples mainly from Great Britain and New Zealand can be found in Packham & Willis (1997, pp. 230-235).

8.3.2 Plant community variation in Europe

The presence of perennial species provides a distinctive geographical range of variation. In the north the oyster-plant *Mertensia maritima* is a rare but characteristic plant of semi-stable shingle beaches and may be accompanied by *Ligusticum scoticum*. These communities develop on exposed shores where periodic storms may displace the shingle and along with it the plant communities. Recolonisation takes several years as stability returns.

Gravel beaches have been recorded in the High Arctic on Svalbard with scattered plants of *Cerastium alpinum*, *Cochlearia officinalis* ssp. *groenlandica* and *Sagina intermedia* (Dobbs 1939). On the coarse shingle beaches in estuaries of the low Arctic (Iceland, North Norway, Russia) *Mertensia maritima* is the only common species (Fjelland 1983). Where sand is mixed with the shingle *Honckenya peploides* is always present and is replaced inland by *Polygonum oxyspermum* ssp. *raii*. Further south *Eryngium maritimum* and *Beta vulgaris* exist in a matrix of larger pebbles.

Lathyrus japonicus occurs where the shore is relatively stable with a higher proportion of sand. These communities are especially well represented in East Anglia.

The shingle vegetation in Great Britain includes pioneer communities found adjacent to the sea and subject to the greatest maritime influences. Secondary pioneer communities are found further above mean high water mark (Figure 8.63). Grassland communities, which may reflect differing degrees of maturity and substrate influences, are also found on shingle. Heath communities, indicative of both wet and dry conditions, are primarily confined to northern sites such as at Kingston, Grampian, northeast Scotland, where they occur in the unafforested areas of both disturbed and undisturbed shingle. Scrub communities are common on shingle and in a few cases there is seral development to woodland (Randall 1996).

Shingle beaches are common around Ireland often with luxuriant vegetation. Foreshore communities include *Beta vulgaris* ssp. *maritima*, *Crambe maritima*, *Lavatera arborea* and *Tripleurospermum maritimum* similar to British western and south-western shingle. In the north of Ireland from Donegal to Down, *Mertensia maritima* communities are recorded (Farrell & Randall 1992).

France also has a good selection of shingle sites. Particularly rich botanically is the pebble area at Sillon de Talbert on the north Brittany coast (Géhu 1960). This site has lows with *Atriplex portulacoides* and *Rumex crispus* amongst a number of pioneer species similar to Essex and Suffolk shingle. The Armorican gravel ridges further east are noted for coastal ecotypes of *Erodium cicutarium* ssp. *pinnatum*, *Galium neglectum*, *Galium verum* var. *maritimum* and *Jasione crispa* ssp. *maritima*. A typical community found in southern England, Belgium and France is described in Table 8.21 and shown in Figure 8.63.

Table 8.21. Plant communities of shingle beaches on the Atlantic coast of Belgium and France (Géhu & Géhu-Franck 1993)

Zone	Plant communities	Main species
Outermost ridge	Atriplicetum glabriusculae	*Atriplex glabriuscula, Crambe maritima & Glaucium flavum*
	Lathyro-Crambetum	*Rumex crispus*
2nd and 3rd ridge	Trifolium scabrum community	*Trifolium scabrum, Sedum acre, Desmazeria marina & Sagina maritima*
Interior ridge	Ulex-Polypodium community	*Ulex europaeus* ssp. *europaeus & Polypodium vulgare*
	Arrhenatherum community	*Arrhenatherum elatius & Silene vulgaris* ssp. *uniflora*

Figure 8.63. Shingle communities; *Rumex crispus* and *Glaucium flavum*, southeast England

In the Mediterranean a number of species common on unstable beaches are present. Vegetation also occurs on minor beaches of coarse cobbles on the coasts and islands of Croatia (Lovric 1993a).

8.3.3 Rare plants

Rare plants on shingle are few in number but important, as they represent several geographical variations. *Mertensia maritima*, a northern species, is one of the more celebrated plants of shingle habitats in Europe. It is both a specialist of semi-stable shingle beaches and an indicator of change, being one of a small number of species which have retreated northwards in recent years. *Lathyrus japonicus* is another species which was formerly distributed as far south as Cayeux sur Mer. This site was the last recorded location before it became extinct in France. Apart from the loss of species, these occurrences reflect the nature of use of the shorelines, with the retreat of colonies at many sites appearing to be linked to human disturbance and climate change (Farrell & Randall 1992). Stony beaches in Greece also have an interesting flora (Polunin & Walters 1985) with widespread, though local species:

Cakile maritima
Salsola kali
Crithmum maritimum
Eryngium maritimum
Limonium vulgare
and species with a more southerly distribution e.g. :
Camphorosma monspeliaca
Atriplex pedunculata
Salsola soda
Arthrocnemum fruticosum
Arthrocnemum glaucum
Matthiola tricuspidata
Eryngium creticum
Limonium cancellatum
Inula crithmoides and
Artemisia caerulescens

8.3.4 Invertebrates

Coastal shingle is of high interest for terrestrial invertebrates where it extends out of reach of normal high tides and has some vegetation. Generally it presents a hostile environment, being very dry with great extremes of temperature. These same extremes can provide, on the other hand, warmer conditions than the surrounding habitats, especially in northern areas and can lead to an enriched fauna. The specialist plants of the unstable early stages of colonisation have their own fauna. The rare weevil *Ethelcus verrucatus* is found on the roots of *Glaucium flavum* in Great Britain. *Cytisus scoparius* and *Prunus spinosa* are important species of shrub supporting both specific and more generalist invertebrate species. Extensive shingle banks can have an extremely rich fauna and the presence of many bees, wasps and ant species which rely on sparsely vegetated or open ground is noteworthy. Of special interest amongst these groups are bumble bees with some individual sites being amongst the most important for this group of species. For example Dungeness has more bumble bees (13 out of 17 *Bombus* species) than any other locality in Great Britain (Williams 1989) and a rich invertebrate fauna generally (Morris & Parsons 1993).

Other rarities include the scaly cricket (*Pseudomogoplistes squamiger*), a Mediterranean species discovered only on the south coast of Britain in 1949. This species is specifically associated with seaweed-strewn strand lines on shingle beaches, where it feeds at night (Sutton 1999). A number of other rare species found on shingle 7 spiders, 2 wasps, 7 beetles and 2 flies are restricted to the habitat in Great Britain, whilst being more frequent and widespread in mainland Europe on open heaths and grasslands (Shardlow in Press).

8.3.5 Other animals

Shingle beaches and structures provide suitable habitat for a small number of other species. These are generally ground nesting birds which make a simple scrape where they lay their camouflaged eggs. Some of the main European species using shingle shorelines for breeding are listed below.

*Haematopus ostralegus** (oystercatcher)
Larus canus (common gull)
Sterna caspia (caspian tern)
*Sterna albifrons** (little tern)
Sterna hirundo (common tern)
Sterna paradisaea (Arctic tern)
*Charadrius hiaticula** (ringed plover)
Oenanthe oenanthe (wheatear)

* Species most commonly breeding on shingle shores.

Turnstone (*Arenaria interpres*), oystercatcher and redshank are found outside the breeding season feeding along the strandline on rocky/shingle shores. Stone curlew (*Burhinus oedicnemus*), a species of open sandy soils also nests on shingle. Dungeness was a stronghold for the species in Great Britain and nested in some numbers on its "stony wastes". It was still nesting in small numbers in 1950 but no longer breeds there. Their demise is thought to be related to predation by carrion crows, though egg collectors in the late 1800s accounted for many of the birds (Harrison 1953). The open expanses of larger sites may also be important for hare (*Lepus capensis*) and fox. The latter can take a major toll on the eggs and young of ground-nesting birds.

8.4 Human activities and conservation

Shingle beaches by their nature tend to be unstable and prone to sometimes massive and rapid change. The plants and animals which survive there or in the early stages of the development of shingle structures are therefore usually tolerant of periodic disturbance. However once the ridges become stabilised and out of reach of storm waves, the gradual build up of interstitial sediment takes place and with it the development of vegetation. At this stage the communities which become established are extremely sensitive to disturbance. The most widespread and damaging activity is the removal of gravel either from the beach or more significantly the structure itself. This destroys the vegetation and associated fauna and because of the complex nature of the matrix in which they occur, they are difficult to recreate. The gravel pits (mostly flooded) which often develop following extraction create a different habitat which may or may not substitute for the loss of interest (Chapter 9). Infrastructure development can be equally damaging and the simple act of driving a vehicle across a series of shingle ridges may create a scar which can remain for many years, until the next 'natural' change takes place and the structure is reformed.

Other management issues of significance are the impact of building and maintaining sea defences which can cause major change to the sediment regime. Knock-on effects may result in erosion elsewhere and/or the need for further protection. Reprofiling of the beach to 'improve' its sea defence capability is a normal practice, but this also destroys any incipient vegetation. Recreational activities are important, affecting both shingle beaches and structures. Grazing is a much less of an issue than for other sedimentary coastal habitats. As with other habitats, their direct loss through human action, combined with sea level rise and the impact of increased storm frequency and intensity, can narrow the zone considerably. Even quite small perturbations under these circumstances cause lasting and irreparable damage to this normally robust system. A summary of the most significant activities affecting shingle beaches and shingle structures is shown in Table 8.22.

Table 8.22. Key management issues affecting shingle areas in Europe

HABITAT LOSS THROUGH:	RECREATIONAL USE:
Gravel extraction;	Trampling;
Infrastructure development, industry &	Disturbance (nesting birds);
housing;	Disturbance (vegetation);
Car parks, airfields and roads;	Boat mooring (including fishing);
Power stations (including nuclear).	OTHER ISSUES:
SEA DEFENCE & COASTAL PROTECTION:	Grazing (loss of rare plants);
Groynes and artificial breakwaters;	Water abstraction;
Reprofiling fringing beaches;	Military training;
Sea level change.	Vehicle access.

8.4.1 Habitat loss

A distinction can be made between gravel extraction and infrastructure development. Both destroy the surface shingle and its nature conservation value (Figure 8.64). However, the former leaves a void which can be recolonised, or when filled with water potentially provides alternative habitat; the latter does not. Other uses are less obviously destructive to beaches, such as hauling out fishing boats and fishing gear or recreational activity.

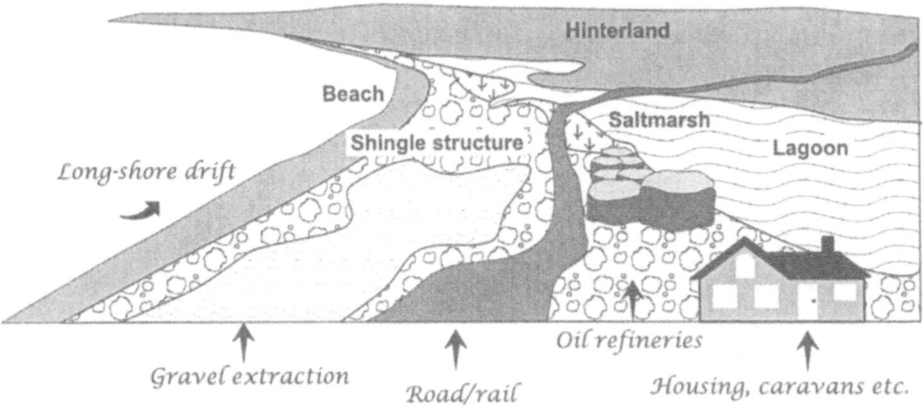

Figure 8.64. Some damaging uses of shingle beaches & structures

Shingle structures and to a lesser extent shorelines provide a source of easily accessible material for the construction industry. Several large sites have been adversely affected by gravel extraction. Dungeness, the largest shingle structure in Great Britain with 1,700ha of exposed surface, is the most disturbed site with some 43% of its surface affected by a wide variety of adverse impacts; the most significant of these is the complete destruction of the vegetation and ridge structure through gravel extraction (Fuller 1985, Chapter 9). A similar and even more complete loss has overtaken the Crumbles in southern England. The natural shingle surface of this 170ha site immediately east of Eastbourne, has virtually been destroyed in recent years by building works and gravel extraction. The remaining vegetation on intact ridges has been broken up by heavy visitor pressure.

A similar story can be told for many other areas. For example much of the surface shingle west of Rye Harbour has also been destroyed by gravel extraction. The Kingston Shingle which lies to the west of the River Spey in Scotland is another important area where shingle extraction has destroyed much of the ridge structure, probably during the 1940s. Two sites in France, Cayeux sur Mer and Sillon de Talbert, have both been subjected to extraction.

On the eastern shore of Nova Scotia, Canada, extraction of gravel shorelines for the construction industry has taken place. In particular Taylor et al. (1985) report the effects of excavation of shingle between 1954 and 1971 when half of the structure (2 million tons of sediment) was removed. This resulted in the total collapse of the barrier and the rapid landward migration of the area that remained.

There are few examples of major and extensive infrastructure development on shingle structures, though houses have been built on

Dungeness, the Crumbles and at Slapton (Figure 8.65) in the UK. Dungeness also has an airport, nuclear power station and an assortment of other buildings including a lighthouse, bird observatory and public houses. Roads and a railway complete the mosaic of surface shingle loss. At Orfordness on the east coast of England, the building of a wartime radar station and atomic weapons research establishment has also destroyed the surface vegetation to the north of a National Nature Reserve. Though less information is available, the French sites appear to have been similarly affected; for example "The small sand and gravel ridges of Basse Normandie (south of the Seine) have in large part been urbanised." (Géhu & Géhu-Franck 1993).

Figure 8.65. Slapton shingle beach, Dorset. A road, buildings, car parks and sea defences affect the shingle bar which separates a lagoon from the sea

8.4.2 Sea defence and sea level rise

Shingle sites serve important sea defence functions, a role which has long been recognised. Discussions over shingle extraction during the 1911 "Royal Commission on Coastal Erosion and the reclamation of tidal lands in the United Kingdom" prompted many exchanges including the following:

"Do you recommend that the removal of shingle, whether for manufacture of concrete, road-making, or ship ballast should be stopped?"

"I think that any beach that can be shown in any way to protect the coast should be left alone .... I think that ought to be enforced very strongly indeed. In many cases shingle is taken from comparatively narrow, small masses of beach, the decrease of which leads to very serious results, that is to say, the damage done is many times the worth of the shingle taken". [Vol. 1(2); page 92; minute 2278] (Royal Commission on Coastal Erosion 1911).

The importance of this comment is amply borne out by the fate of the Hallsands (Figure 8.66a) fishing village in south Devon. In 1897 work began on the construction of Devonport docks for the Royal Navy. Large quantities of stone were needed which appeared to be present in abundance offshore in nearby Start Bay. Because of the potential effects to sea defences protecting nearby land, the terms of the licence included a clause allowing the licence to be revoked at short notice. By 1900 the villagers of Hallsands noticed the beach which protected the houses had begun to disappear. Damage to the houses resulted in the licence eventually being cancelled in 1902. By then the damage was done and all that remains of the village today are a few houses perched precariously above the eroding shore (Figure 8.66b). Access to the area is now completely restricted.

Figure 8.66a. Hallsands village in 1894 from a postcard by Valentine & Sons. The beach protects the village from erosion

Figure 8.66b. Hallsands 'village' in 1994 following erosion of the beach. The house in the foreground (still inhabited in 1994) is arrowed in the previous postcard picture

It is not absolutely certain that the extraction caused the loss of the beach as changes to wave patterns and the incidence of storms can also be implicated. Perhaps the beach waxed and waned naturally. However, what does seem to be clear is that with no new sediment to replace the beach from offshore resources, the village was doomed. This example suggests that where there are shingle deposits great care should be taken when interfering with the system in which they occur. Not heeding these warnings can have long term implications that are damaging to wildlife and also to the sustainability of human use. This may be especially true during a period when global sea level is rising.

8.4.3 Recreation, agriculture and forestry

On the most mature shingle structures even relatively small incursions into the surface layer may remain visible for many years. The impact of off-road vehicles range from relatively minor damage to intact vegetated ridges which break up the vegetation, to prolonged use and total destruction. Off-road vehicles can impact on some of the seemingly most inaccessible areas. The spit of Orfordness National Nature Reserve, formally accessible only by boat, can now be reached by motor cycles and the like from Aldeburgh. Tracks can now be seen near the end of the spit, many kilometres from the access point, where none has appeared before.

Agricultural use does not have a major influence on shingle areas though low-level grazing may take place. Rhunahaorine Point in Argyle, Scotland is a relatively small site which consists of a series of ridges covered with dry and wet heath vegetation. This site represents a very different set of management issues. The use of artificial fertiliser, reseeding and increased stock levels have reduced the area of heathland, particularly on the shingle ridges. Destruction through the extraction of gravel has not occurred, and the structure remains intact.

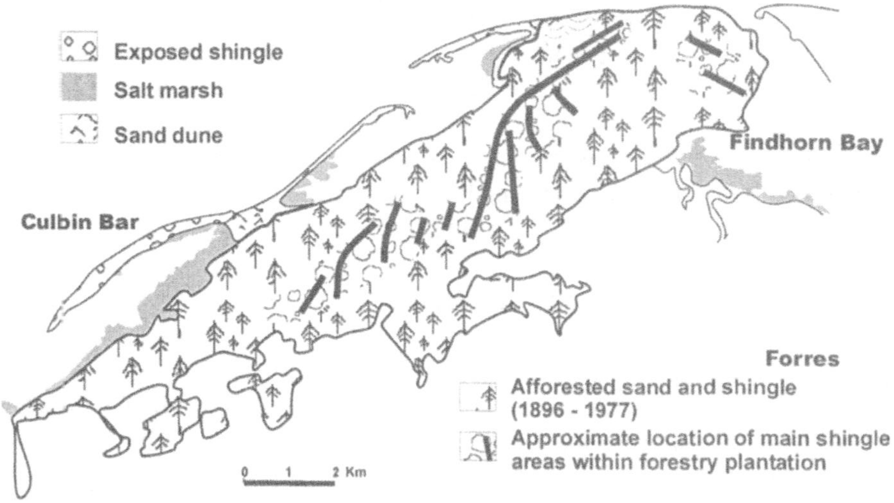

Figure 8.67. Afforestation of Culbin Sands and shingle in relation to the Culbin shingle bar

Shingle structures are usually too dry at the surface to support anything but rudimentary woodland and for this reason afforestation is not a major issue. One exception is the shingle system between Culbin and Kingston in northern Scotland. Culbin shingle bar, which lies off the north coast of Moray in Scotland, is one of the least affected by human activity. Access to the offshore bar is difficult and the site shows a good series of shingle ridges running parallel to the coast. This forms part of a much larger complex of shingle ridges, sand dunes and saltmarshes (Hansom 1999). Within this area there is complex of shingle ridges and sand dunes described by Ogilvie (1923) when many of the natural features could still be seen. Afforestation of the sand dunes, which cover much of the underlying shingle, is considered in chapter 7. However, it is clear that, although not so well documented, the majority of the mainland shingle ridges have also been planted with conifers. Although areas of intact bare, and naturally vegetated, parallel shingle ridges can still be seen, these are very restricted (Figure 8.67 above).

8.4.4 Water resources

Shingle beaches and structures are often associated with lagoons. Indeed as, Chapter 10 shows, many important coastal wetlands are dependent on their relationship with enclosing protective spits, bars and barriers composed of sand and/or shingle. In the case of Dungeness small pits created in natural depressions on the surface support important freshwater plant and animal communities (Waters 1985, Waters & Ferry 1989). Because of its size, the water stored beneath the surface represents a resource in its own right. Water abstraction for drinking purposes has lowered the water table. This has led to a reduction in open water as marginal vegetation has closed in and threatens the ultimate loss of conservation interest in these small but unique habitats. The process can also promote saline intrusion which not only changes the species composition but also can have a equally important impact on the suitability of the water used for drinking.

8.5 Options and opportunities

Destruction of shingle beaches and structures is widespread. On major shingle structures this has often been at the expense of valuable flora and fauna. Managing those areas that remain is in theory a relatively easy task as prevention of damaging activities is all that is required. Too often, however, other economic incentives force change. In these cases, what is also overlooked is the inherent importance of many of these areas (beaches and structures) for sea defence. The fact that beaches provide effective protection from the sea especially during storms is an important additional incentive for retaining all those that remain. Interference by human action not only destroys their inherent conservation value but can lead to irreversible and costly threats to human life and habitation. Recognising the importance of both of these is essential for assessing the options and opportunities for effective management.

8.5.1 Managing shingle banks for their flora

Some of the rarer plant communities on shingle are unique to this substratum and highly susceptible to disturbance. Slow-growing bryophyte and lichen-rich communities develop on the thin soils. These communities tend to be present in remote areas with limited public access (e.g. Orfordness, lee slopes of Chesil Beach in the UK). To facilitate the continued development and survival of these lichen or moss-rich heaths, access should be restricted. Trampling disturbance is a particular problem with these and some pioneer herb communities (Scott & Randall 1976, Randall 1977). Here, as with direct destruction of the shingle surface by excavation or building, avoidance of the damaging activity is the best conservation option.

For other types of vegetation varying degrees of management intervention are required. Minimum management is needed to maintain scrub or woodland communities. But heath, grassland and pioneer communities may require more active management. *Calluna vulgaris* heath generally requires low-level grazing to maintain age and species diversity. However, edaphic and environmental conditions peculiar to shingle seem to serve as natural limits on cover for *Calluna* in locations such as the Culbin Bar where hares may be the only grazing animal. Active management may be needed to restrict scrub invasion at other areas but there is little evidence for this from the major sites in Great Britain. Away from the drift-line, shingle structures are usually highly nutrient-poor so that the introduction of grazing herbivores will almost always cause vegetation change. Locally, though, there may be enrichment from nesting birds which will influence the vegetation composition such as is increasingly the case at Orfordness where there has been a recent rapid rise in nesting gulls (Cadbury & Ausden in Press). A similar effect has been recognised from the Nelson barrier, New Zealand (Randall 1992)

The retreat northwards of *Mertensia maritima* and the general reduction in range of *Lathyrus japonicus* (Randall 1987) and *Crambe maritima* prior to 1930, all species which are restricted to shingle shores, are of interest. The reasons for this change are not clear though increased recreational use of the beaches seems to be important. Colonies of *Mertensia maritima* present along High Water Mark Ordinary Spring Tides may disappear during severe winter storms. Grazing may also reduce the viability of colonies since the flowers are especially sought by sheep and rabbits (Randall 1987). Climate change may also be a factor (Farrell & Randall 1992). The extinction of *Lathyrus japonicus* from the last French site appears to be due to a variety of factors including grazing, tourism and shingle extraction.

8.5.2 Sea defence and nature conservation

Shingle sites are essentially mobile structures developed in dynamic high energy environments. This inherent instability presents a major potential problem for large constructions built on shingle areas, or areas protected by shingle banks. Natural events will cause erosion and the movement of material. Moreover where there is interference with natural processes, either by removing shingle from the structure itself or from areas providing source material, its stability can be threatened. Thus in addition to the destruction of the surface vegetation, it may also put at risk any structures that have been built on or behind the shingle. Low-lying land protected by shingle beaches or structures may be particularly vulnerable to flooding.

Orfordness provides a case study allowing an examination of coastal protection policy in the United Kingdom. The evolution of the coastline here is relatively well understood (Steers 1927, Carr 1970, Fuller & Randall 1988). For centuries the natural movement of beach material by long-shore drift has resulted in a gradual progression of the Ness southwards. Inspection of early maps, dating from 1530, show that it has moved from just north of east of the village of Orford to due east. At the same time the spit has grown rapidly southwards (Anon 1977, Figure 8.68 above) but receded somewhat since 1893. In its long, several thousand year history, the narrow ridge which has separated the Rive Alde from the sea south of Aldeburgh has remained intact, at least in the medium to long term. However, despite this, concern about flooding resulted in a sea wall being built to protect the town of Aldeburgh. This wall was extended into the northern end of the Ness just before a major storm surge which affected the southern North Sea in 1953. Where the sea wall stops there is now a thin and apparently vulnerable sea defence which has been reinforced with imported shingle.

Because of the threat of flooding to properties in Aldeburgh and the agricultural land lying below the 5 metre contour in the valley of the River Alde, a cost benefit analysis was carried out for the Ministry of Agriculture, Fisheries and Food (the grant-aiding body), by the University of East Anglia (Turner et al. 1990). This concluded that the benefits associated with the avoidance of agricultural loss, property and environmental damage and heritage/recreational losses were in excess of the costs of a full sea defence scheme. In reaching this conclusion the conservation value of the area was based on advice which assumed that the existing interest was such that it should be protected. The presence of a number of statutory protected sites, nature reserves and areas of high landscape quality reinforced this view.

The evolution of the structure of Orfordness has been described in a number of papers, and its dynamic nature stressed (Fuller & Randall 1988). Despite this, its natural evolution was not included in the costed options associated with defining the sea defence strategies. It is perhaps understandable, given the historical perspective referred to above, that this should be the position adopted by those advising on the environmental considerations. Not only does it confirm the value attached to the existing habitats and landscape but also, given the uncertainties of the results of a major breach, it has a reasonably predictable outcome. It also reinforces the traditional engineering view associated with maintaining the existing line of defence. However, looked at from a different perspective, which accepts change as a natural and perhaps healing force in coastal situations, the cost-benefit ratio might have been different.

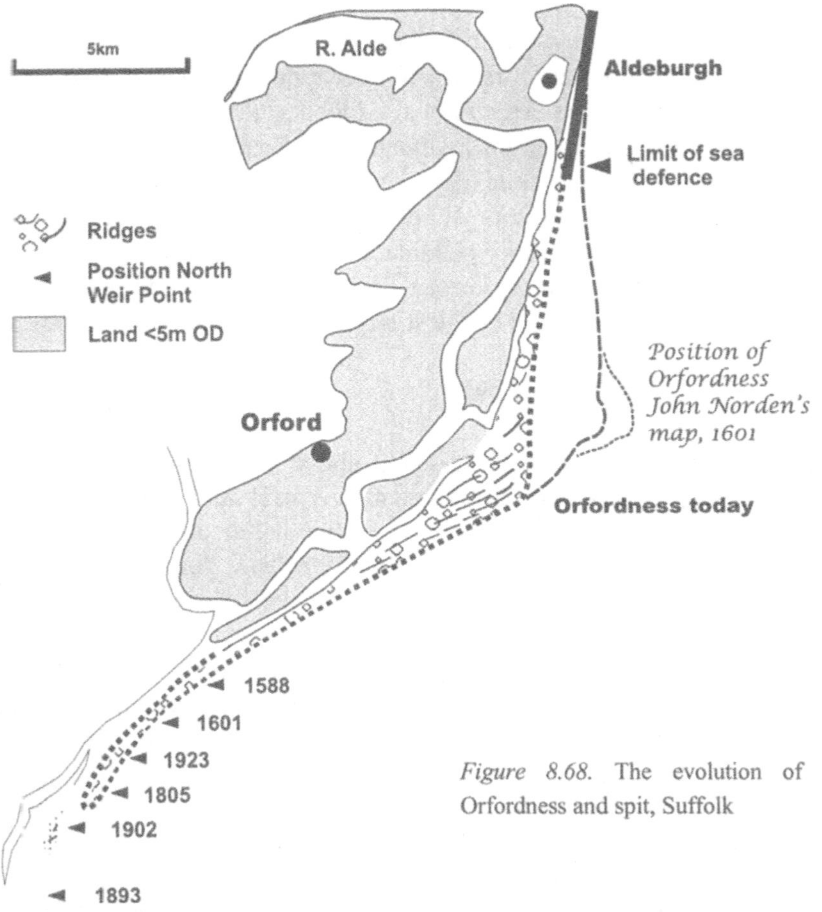

Figure 8.68. The evolution of Orfordness and spit, Suffolk

Orfordness has much in common with Dungeness in Kent as it existed some 700 years ago, just before the River Rother broke through at Winchelsea (Fuller & Randall 1988, Chapter 11). At that time the spit had forced the river 20km southwards along the coast much as Orfordness had deflected the Alde. They speculate that a breach in the vicinity of the Slaughden wall might allow the beach to develop naturally, and could provide a more cost-effective solution to maintaining the sea defence.

At the northern end of the site immediately south of any breach, it is likely that the Ness itself would begin to erode and continue its journey southwards. This would in fact have a double benefit for nature conservation. Firstly it would help re-establish the geomorphological processes, and allow the regeneration of a natural shingle surface in an area which has been adversely affected by human activities. Secondly the possibility of increased flooding of agricultural land along the River Alde might result in the recreation of mudflats or saltmarsh behind the existing sea wall. In some areas, like Havergate Island (a nature reserve), this might mean the loss of one wildlife habitat for the gain of another. Given the possible overall benefits for nature conservation, this might be acceptable. In carrying out the original evaluation at Aldeburgh, both the sea defence needs and those of the environment assumed that maintaining the status quo was the desired objective. Thus the traditional view of the engineer was reinforced by the protectionist philosophy of the conservationist. As environmental considerations become more important when designing and costing engineering options for sea defence schemes, alternative strategies which accept change as part of the strategy must be included (Doody 1992).

8.5.3 New habitats for old

Gravel extraction has been considered above because of its destruction of surface shingle, together with the unique plant and animal communities which occur. At Dungeness (Chapter 9) this has had a devastating impact, destroying large areas of vegetated shingle ridges. The extent to which the newly opened surface is recolonised depends on the fine fractions within the surface and the depth of the excavation. Deep gravel pits often provide open water sites suitable for water birds and because of their coastal location can develop significant populations. From a nature conservation point of view there may on the face of it be little difference between a coastal gravel pit and excavations elsewhere. However, artificial wetlands such as gravel pits quickly develop their own conservation interest, which with appropriate management can become quite significant. It is debatable whether the loss of rare plant and animal communities on stable shingle can be equated with the gain of wetland bird populations, found in gravel pits elsewhere (Chapter 9).

When deep excavations are interspersed with shallower areas, which remain damp but not wet, some of the former interest may become re-established. Shingle heath has developed in excavations on the Kingston shingle in Scotland and includes the rare orchid *Corallorhiza trifida*. The Rye Harbour shingle complex, also extensively excavated, includes a range of wetlands and damp marginal habitats with a rich invertebrate fauna. Perhaps it is a matter of personal preference as to whether the loss of one habitat can be adequately replaced by another. Given the rarity of stable, natural shingle banks and the specialist nature of many of the species, conservationists should err on the side of caution and further losses of habitat should be avoided. Because of the importance of shingle beaches and structures to coastal defence more thought must be given to combining the nature conservation value with that of sea defence. The structures should be seen for what they are: a natural dynamic cost-effective sea defence, often including high nature conservation value, rather than as just another exploitable resource.

9. NATURE CONSERVATION AT DUNGENESS - A CASE HISTORY

9.1 Introduction

Dungeness (Figure 9.69) is one of the largest shingle cuspate foreland structures in the world, and is a worthy subject for more detailed discussion of the relationship between biodiversity and integrated coastal management. A well-studied geomorphological system with extensive and intensive human activity affecting the surface shingle, its future is largely determined by local approaches to planning and shoreline management. Despite the adverse treatment it has received, touched upon in Chapter 8, it also remains one of the most important nature conservation sites in Europe and has been the subject of detailed scientific study (Ferry & Waters 1985, Ferry et al. 1989, Ferry in Press).

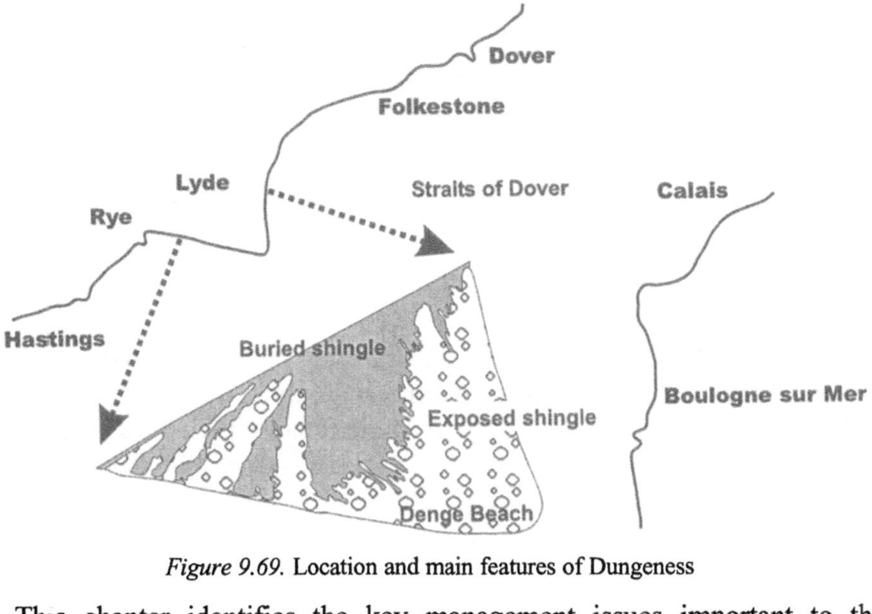

Figure 9.69. Location and main features of Dungeness

This chapter identifies the key management issues important to the conservation of the site, examining the way in which recent developments in coastal management have been applied to both the site and the geomorphological unit in which it lies. Traditionally the descriptions and scientific studies for nature conservation focus on the present cuspate foreland. This is relatively recent in origin and the most active part of a much older and larger system stretching from the cliffs east of Hastings in the west as far as Folkestone in the east.

9.2 Evolution of Dungeness

The present structures are the result of geomorphological processes which have progressed over a period of 4,000 years or more, since sea level stabilised at about its present position. The sequence of growth is described and interpreted by a number of authors notably Steers (1969a, pp. 318 - 331) and summarised in Table 9.23 and Figure 9.70 below.

Table 9.23. Important dates in the geomorphological evolution of Dungeness

Date: years BP	Situation
12,000	Peat can be found at -25m below present mean sea level at Tillingham
Foreland initiation 4,000- 6,000	Sea level rising but less rapidly and beginning to stabilise about its present level. Formation of barrier.
Foreland stability 2,000 - 4,000	Sand flat replaced by silty clay then by peat. The oldest sediments formed 6,000 years BP. Peat spread over Romney Marsh in a protected environment behind the barrier. Pulses of gravel produced by storms and climate change; sea level more or less stable.
Breakdown 2,000	Reduction in sediment supply. Youngest peat deposits formed when inundation by brackish/marine conditions followed the breakdown of the barrier/spit, then enclosing the marsh.
Roman to present	Relatively low sea level in 'Roman times' allowed enclosure. Rise in sea level to the present day.
800	Storms.
From 700	Storm ridges build out at Dungeness.

Between 17,000 years BP and 'Roman times' as the land rose (and fell) and sea levels changed, a series of spits and bars enclosed the area which today is known as Romney Marsh. During these periods the area was covered by swamps and forests. In more recent times 'the Roman Uplift' recovery of the land surface allowed the northern part of Romney Marsh to be enclosed and used for agriculture. At the same time a bar stretched from west to east across the bay and the river Rother probably exited to the sea much further east than at the present day (Figure 9.71).

Most of the change to produce its present form took place after 1287 when a sequence of violent storms breached the shingle bar, which has since been eroded, and the River Rother changed its course to exit near its current location at Rye. The present Ness represents a series of new beach ridges which grew in a south-easterly direction in earlier times, at least 700 BP.

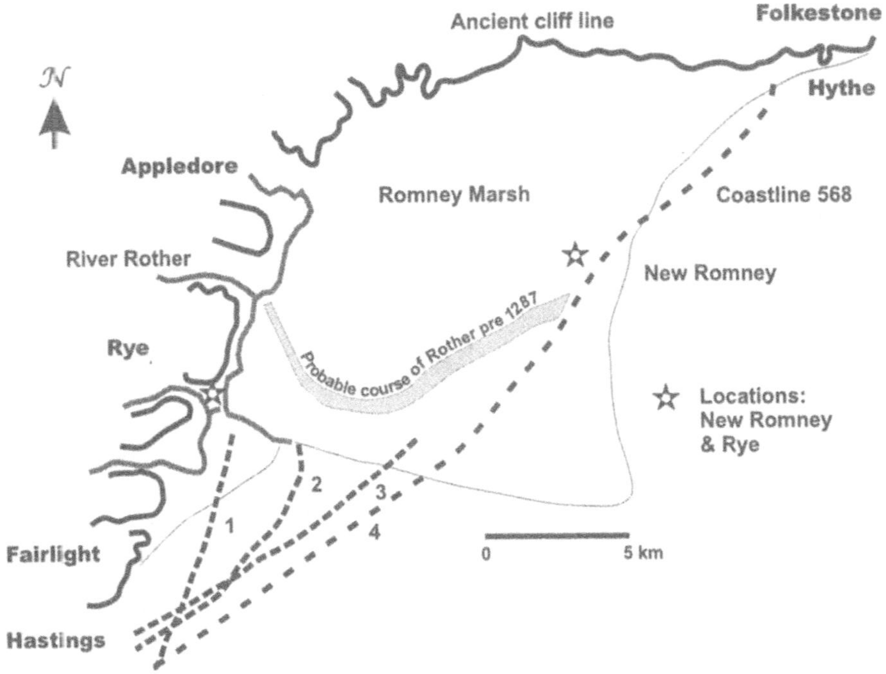

Figure 9.70. The development of the shingle bars and spit. Ridges 1 - 4 represent a sequence up to and including 'Roman times' when Hythe was probably the chief outlet to the sea

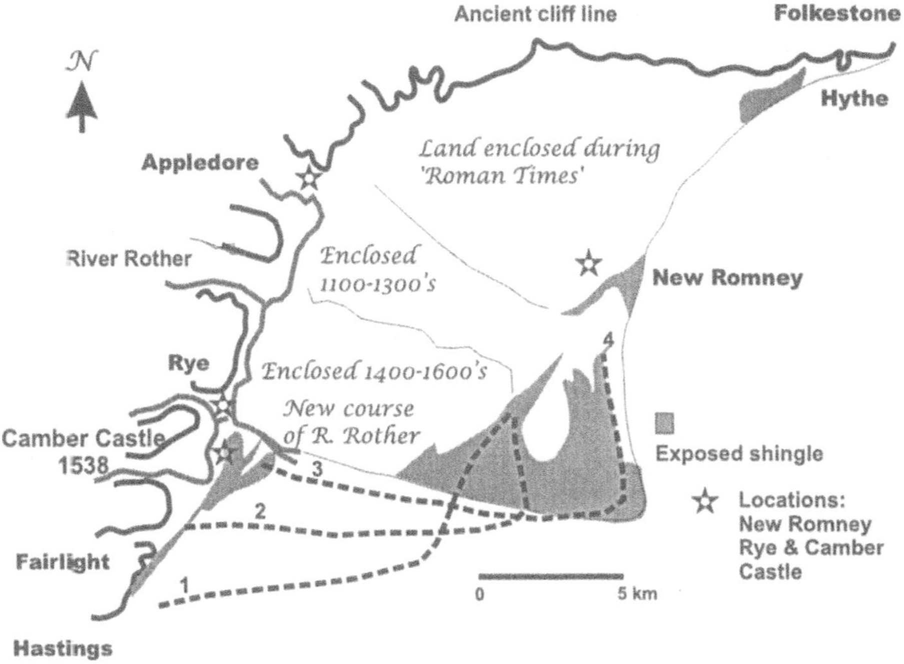

Figure 9.71. Development of the cuspate foreland of Dungeness since 'Roman times'

9.3 Nature conservation importance

From the above it can be seen that the Dungeness foreland represents a relatively recent stage in the evolution of a shingle system stretching along the coast from west of Rye to Hythe. It is the foreland which supports the significant biological features described below. In common with other shingle systems, the older ridges became stable as they were removed from the influence of the sea. The accumulation of fine material in the spaces between the pebbles allowed thin soils to develop, and with it the establishment of vegetation. The vegetation succession is shown in Table 9.24 after Ferry et al. (1990).

This sequence of plant communities and their natural succession on the surface shingle is older and more diverse than at other shingle sites in Britain. Added to the maturity of the vegetation there are also natural cycles of degeneration and regeneration particularly of the shrub *Cytisus scoparius* (Packham & Willis 1997, Figures 8.5 & 8.6). As will be discussed below these cycles add greatly to the biological diversity of the site, particularly in relation to the invertebrate fauna.

Table 9.24. The main types of shingle ridge vegetation present at Dungeness

Coastal ridges	**Wetlands** (Waters & Ferry, 1989)
• First strandline ridge, scattered *Atriplex glabriuscula*	• Emergent vegetation with *Typha angustifolia* or *Phragmites australis*
• Strandline vegetation dominated by *Crambe maritima*	• Sallow carr with *Salix cinerea* and a number of ferns (originally saline percolation lagoons)
• Mesotrophic grassland dominated by *Arrhenatherum elatius*	**Older inland ridges**
• Scrub community dominated by *Cytisus scoparius*	• Calcifuge grassland with *Festuca filiformis* dominant and lichens
Eroded and exposed southern ridges	• Scrub with *Sambucus nigra* and *Prunus spinosa*
• Maritime communities with *Silene maritima, Armeria maritima* and the lichen *Cladonia rangiformis* where erosion has re-exposed them	• The 'holly wood'

9.3.1 Rare plants

The site supports a number of rare plants, including species nationally rare in Great Britain with <150 10x10 km squares in the Atlas of the British Flora, Perring & Walters 1976). These include the following:

- *Crepis foetida* reintroduced (in the 1990s) following extinction in 1980 from Dungeness via seed banked at Kew Gardens. Also rare in Europe on shingle beaches in France and Belgium;
- *Silene nutans* and several small clovers e.g. *Trifolium suffocatum*;
- Lichens very important including *Cladonia mitis* which is found only at Dungeness and four other sites in Scotland, North Wales and southeast England. Other species are *C. bacillaris* and *C. ciliata* var. *ciliata*;
- Mosses including *Antitrichia curtipendula*;
- *Carex disticha* sudden loss in 1970 due to rapid fall in water table affecting the Open Pits wetlands, though still present near Lydd Airport.

9.3.2 Invertebrates

The invertebrate fauna of Dungeness is very rich in a wide variety of rare and common species due to the open, dry and warm nature of the site, the absence of intensive agriculture and the rich and diverse vegetation. Ridges with *Cytisus scoparius* (Figure 9.72) are amongst the best habitats because of their structure, height and the hot micro-climate. The list which follows (Table 9.25) provides a summary of some of the more important species associated with shingle and shingle vegetation (Philp & McLean 1985, Morris & Parsons 1993).

Figure 9.72. Cytisus scoparius on mature shingle ridges; an important habitat for rare invertebrates at Dungeness. Note the Power Station in the background

Table 9.25. The principal invertebrate species of special conservation value at Dungeness

Lepidoptera	**Host plant/s**
Lasiocampa trifolii ssp. *flava* (Grass eggar)	*Cytisus scoparius*
Thalera fimbrialis (Sussex emerald moth)	*Achillea millefolium* and *Ulex europaeus*
Dasychira fascelina (Dark tussock)	*Cytisus scoparius*
Eilema pygmaeola ssp. *pallifrons* (Pygmy footman)	Lichens
Hadena albimacula (White spot moth)	*Silene nutans*
Calophasia lunula (Toadflax brocade)	*Linaria vulgaris*
Clostera anachoreta (Chocolate tip)	*Salix* spp.
Diptera	**Vegetation**
Stratiomys furcata (Soldier fly)	Umbelliferous flowers
Platypalpus articulatus - a predatory fly	Grassland
Sciapus contristans - a predatory fly	Grassland
Pipizella virens (Hoverfly)	Grassland
Hymenoptera	**Habitat**
Solitary bees & wasps	Dry, sunny and exposed open sandy soils for nesting
Ants	Including dead wood (e.g. old gorse bushes)
Hemiptera (true bugs or plant bugs)	**Food plants**
Monosynamma (3 species)	*Salix* spp.
Odontoscelis dorsalis (a 'tortoise' bug)	*Erodium cicutarium*
Coleoptera	**Habitat/host plant**
Dromius vectensis (a ground beetle)	Under old prostrate *Cytisus scoparius*
Anotylus insecatus (Rove beetle)	Rabbit burrows
Aphthona euphorbiae (a leaf beetle)	*Euphorbia* spp.
Longitarsus exoletus (ditto)	*Echium vulgare*
L. ganglbaueri (ditto)	*Senecio jacobaea*
Epitrix pubescens (ditto)	*Solanum dulcamara*
Cassida hemisphaerica (ditto)	*Silene nutans*
C. nobilis (ditto)	*Spergula arvensis*
Smicronyx coecus (a weevil)	*Cuscuta epithymum*
S. jungermanniae (ditto)	*Cuscuta epithymum*
Ceutorhynchus verrucatus (ditto)	*Glaucium flavum*
C. geographicus (ditto)	*Echium vulgare*
C. pumilio (ditto)	*Teesdalia nudicaulis*
Apion affine (ditto)	*Cytisus scoparius*
Hylastinus obscurus (a bark beetle)	*Cytisus scoparius* (dead or dying)
Phloeophthorus rhododactylus (ditto)	*Cytisus scoparius* (dead or dying)
Araneae	Habitat
Apostenus fuscus	Thin soil with grasses, mosses and lichens
Pellenes tripunctatus (a jumping spider)	Shingle ridges

There is a wide range of other species such as rare beetles which inhabit the muddy edge of gravel pits, other wetland species, and those of open sandy soils present on the site. Many of these are associated with habitats created by human use including gravel pits and open sandy areas.

9.3.3 Birds

The site is also important for other animals amongst which the bird populations are given most attention. A number of these are species associated with the natural shingle ridge habitat such as the wheatear and meadow pipit. Rarer species include stone curlew and Kentish plover, neither of which now breeds on the site. The presence of the gravel pits has greatly increased the number of coastal breeding birds including gulls, terns and a number of species of warbler, see below.

9.4 Human activity and conservation

Stable shingle is extremely susceptible to physical disturbance. The main factors causing loss of naturally occurring species are direct destruction of the surface through gravel extraction, erection of buildings and vehicle access. Human interference at Dungeness has been varied in form and considerable in extent. It is considered in Chapter 8; detailed summaries can be found in Findon (1985, 1989) and Fuller (1985). Some of the more significant changes are considered below and in Table 9.26.

Table 9.26. Key issues at Dungeness

SURFACE LOSS	HUMAN ACTIVITY
Gravel extraction causing loss of 420ha (19% of the shingle surface) up to 1976	Water abstraction
Electricity generation, loss of 130ha	General recreational use
Military training affecting 776ha	Fishing
Other infrastructure development: housing; railway; airport; camping and caravan parks	Bird watching
Vehicular disturbance to the surface shingle	Research
	Beach feeding and shoreline protection
	Water sports

Activities, particularly in the former group, have caused a significant and permanent loss of the natural shingle surface together with its vegetation and associated wildlife interest. On any measure of conservation importance Dungeness has suffered permanent loss of a major part of the geomorphological structure of great value on a World scale. Despite these losses there is considerable nature conservation interest which remains and this is reflected in the value attached to the site, including its designation as a statutory National Nature Reserve and as a potential Special Area for Conservation under the European Union Habitat and Species Directive.

9.4.1 A question of Biodiversity

Given the extent of habitat loss and degradation a question arises - 'what is the impact on the biodiversity of the site?'. A few species are known to have become extinct. Amongst the plants *Crepis foetida* was extinct in 1980 and breeding by Kentish plover is thought to have ended following the development of the railway and building of houses (Hill & Makepeace 1989). The extinction of stone curlew, as a breeding species, appears to be part of a wider reduction in its population and cannot be specifically related to habitat loss at Dungeness. Considering the extent of habitat loss at Dungeness it is remarkable that so few species appear to have been totally lost from the site. Indeed it is often the case, notably amongst the invertebrates, that Dungeness is one of the last remaining refuges for species lost elsewhere. In the case of grass eggar moth (*Lasiocampa trifolii* ssp. *flava*) it is "only known in Britain (or the World) from Dungeness" after its extinction elsewhere on the south coast (Philp & McLean 1985).

From the literature it could be argued that rather than reducing biodiversity at the site there has been an increase as a result of human action. For example highly disturbed shingle with an input of fine-grained material along road verges may be much richer in plant species than natural undisturbed grassland on shingle (Ferry in Press). The open waters of the gravel pits have undoubtedly increased the range and numbers of a wide variety of waterfowl. A study of the bird populations between 1969-1987 (the main period when the gravel pits were being excavated) show numbers of waterfowl notably teal, gadwall (*Anas strepera*), wigeon (*A. penelope*), shoveler (*Spatula clypeata*), goldeneye (*Bucephala clangula*) and smew (*Mergus albellus*) were all up dramatically. Breeding birds including common tern (*Sterna hirundo*) and sandwich tern (*S. sandvicensis*) increased as did tufted duck (*Aythya fuligula*), great-crested grebe (*Podiceps cristatus*) and reed warbler (*Acrocephalus scirpaceus*). There was also an overall increase in diversity of passage migrants (Hill & Makepeace 1989).

Thought once to have been widespread in Europe medicinal leach (*Hirudo medicinalis*) appears to have declined due to wetland loss and exploitation. It is known to have only 18 populations in Great Britain and is extinct in Ireland (Wilkin 1989). An important population was recently found in association with a gravel pit at Dungeness and subsequently several other sites have been found. The survival of this relict population may well have depended on the extraction of gravel.

It is of course impossible to make a reasonable comparison between the loss of a rare habitat, such as occurs at Dungeness, and a rise in overall biodiversity. Equally difficult is relating the extinction of Kentish plover or stone curlew as breeding species, with the rise in populations of breeding water birds consequent upon the excavation of the gravel pits. Is the rise in number of species overall a fair exchange for the loss of undisturbed surface shingle and shingle vegetation? Is the fact that the medicinal leach survives here important?

The answer probably differs depending on the interests of the individual making the comparison. Visitors to the Royal Society for the Protection of Birds reserve, which includes a high proportion of areas of excavated shingle, and who see black tern (*Chlidonias niger*) for the first time may consider it to be a fair trade. Those with an interest in the development of 'natural' coastal systems, vegetation and/or invertebrate conservation see it differently. From a purely nature conservation point of view the specialist and more unusual species, which depend on a largely natural feature, are generally more highly rated than the more common species which make up the bulk of the new species. Any coastal gravel pit in southeast England might have been expected to attract a range of birds similar to those present at Dungeness today.

9.4.2 A case for integration

Dungeness today must be viewed as a part, and a relatively small one, of a larger geomorphological system which has been evolving for several thousand years. The Ness has retained much of its wild and open character and the unique wildlife which it supports despite major human use. It is unclear how long this interest can be sustained. Three major considerations apply:

1. Proposals for further development, which will result in the loss of more high quality ridge structure and vegetation, continue to be made. An extension to the airport was proposed, though now dropped, and gravel extraction remains a major activity. Together with other developments, these will cause further destruction of the surface shingle. Water abstraction, military use and power generation also continue and represent both potential allies in the maintenance of biodiversity as well as a continuing threat from peripheral activities such as vehicle access to undisturbed areas. In the past apparently well-intentioned activities including the erection of a series of posts in previously undisturbed shingle "to attract a rare moth" caused considerable damage with little conservation benefit (Findon 1989). This continued chipping away at the remaining undisturbed shingle must eventually have a major impact as the cumulative effects multiply.
2. Southeast England in general, and Kent in particular, are already amongst the most developed parts of Great Britain. With the building of the Channel Tunnel pressures for further expansion of business and an

increase in recreational activity are already manifest. In this context it is already the case that whilst Dungeness was simply a 'wilderness' where only bird watchers and fishermen went, it is now sought out by advertisers and those seeking solace from the pressures of modern life. This will undoubtedly bring a new set of pressures to the area.

3. In the long term, left to the natural forces of sea level rise, tides and storms, the Ness would probably continue to progress eastwards. Eventually this would overwhelm the gravel pits, disturbed areas and artificial structures regenerating the shingle structure as it has done in the past. However, a significant event took place following a Public Enquiry in 1958 which changed the nature of Dungeness, probably forever! Permission was given for a nuclear power station to be built on the Ness close to the sea (Figure 9.72 above). This has effectively resulted in the need to arrest the development of the Ness itself, the most active part of the geomorphological system.

9.4.3 The shingle 'cycle'

Longshore drift is from west to east and the tip of the Ness if left to its own devices would continue to move eastwards as it has done for the last 2000 years or so. The short term consequence of this movement is that there is a deficit of material immediately in front of the nuclear power stations. Clearly this situation is unsustainable and as a result the beach nourishment to build up the shore and protect the building from being undermined and possibly falling into the sea. This is undertaken by collecting material at the east and moving it to the west of the site. Longshore drift ensures the material is carried to the front of the station and back to where it was originally collected in the east, where it is scooped up again and carried back to the west (Figure 9.73).

A further complication lies in the fact that this area of southeast England is sinking and when combined with the general rise in global sea levels there is a relative rise in sea level at Dungeness of about 4mm per year (Barne et al. 1995/8). Given that sea level will probably continue to rise in this area, both because of isostatic change and global sea level rise, then the medium term implications for the sustainability of the power stations poses a real problem.

Failing to take note of natural cycles of evolution can result in a long term and costly need for coastal protection. The question arises as to whether this would have been chosen as the preferred location for such a long term building project as a nuclear power station, had those conducting the original enquiry fully understood the nature of the geomorphological system. So long as the power station is present the Ness must be fixed in its current location, through beach transfer which restricts the range of options for management not only for nature conservation, but also sea defence.

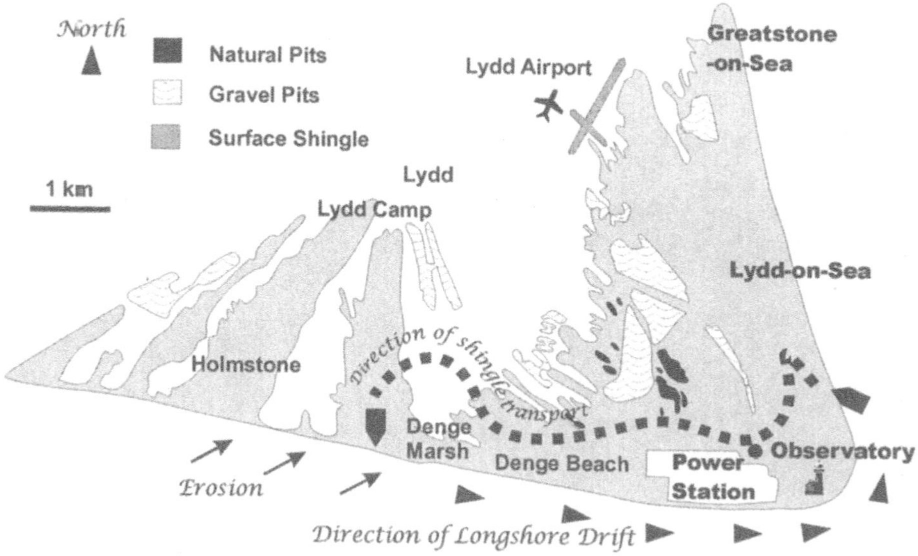

Figure 9.73. Dungeness and the 'shingle cycle', sketch map, after Ferry & Waters (1985)

9.4.4 Lessons

The case of Dungeness illustrates a number of important lessons. Taking our cue from political initiatives, including Agenda 21 from the Earth Summit, then a wider approach to the conservation of the site might be appropriate. Given the geomorphological context in which the shingle foreland at Dungeness has developed and its long evolutionary history, it is concluded that its management should involve the whole of the shingle system stretching from Fairlight in the west, to Hythe in the east and including Romney Marsh! The method of delivering this in the coastal zone is through the mechanism of integrated coastal management, which is considered in Chapter 13. This approach includes an assessment of the resource (including its location, scale, history of development) and its capacity for exploitation. Subsequent development and use is undertaken with both economic and environmental considerations in mind.

The key is to ensure that economic activity is itself sustainable and that in so doing it does forgo the options for the future, in particular in relation to the maintenance of biodiversity. In is clear that in case of Dungeness not only have conservation interests been compromised, but also the sustainability of major infrastructure. Protection of the nuclear power stations is a very long term financial commitment, though the period of electricity generation is a limited one. In the face of rising sea levels this will require more and more resources including gravel transport from the east to the west (Figure 9.74) in order to counteract the natural movement from west to east.

Figure 9.74. Shingle transfer at Dungeness; a never-ending cycle of beach feeding?

10. COASTAL WETLANDS - ESTUARIES, DELTAS & LAGOONS

10.1 Introduction and scope

The term 'coastal wetlands' can be used to embrace estuaries, deltas and lagoons which are here taken together as integrated, complex and dynamic systems. The interactions between the component habitats and species concentrations often mean that one part of the system is dependent upon another. Thus the enclosing sand dune or shingle barrier may help create sheltered conditions for the development of mud flats and saltmarshes to form a 'bar built estuary'. Each component habitat has its own importance which is discussed in the relevant chapters above. This chapter is concerned with those systems which occur in temperate regions and does not include coastal wetlands in tropical and sub-tropical areas (including mangroves).

The nature conservation importance of these areas is most often manifested by migratory birds, notably wildfowl and waders, occurring in large numbers. They also include rare species such as flamingos, pelicans and stilts, for which the habitat is important for breeding and feeding. These and other mobile species (e.g. seals and migratory fish) may use such areas as one of a sequence of geographically dispersed sites important at different phases in their breeding cycle or migration. The habitats also support a range of plants and animals specially adapted to the rigours of the changing salinity, sediment and tidal regimes which characterise these ecosystems. Some species use different parts of the system during different stages in their life cycles. For example resident waders such as redshank may use mudflats for feeding on abundant invertebrate fauna (Figure 10.75), saltmarsh for resting and roosting and adjacent coastal wetlands for breeding. This wealth of interest is built on systems which are complex, diverse and highly productive.

Figure 10.75. Tidal mudflats are rich in invertebrates, the food for large wintering populations of waterfowl visiting estuaries

Early human occupation of coastal wetlands modified their interest rather than destroying it. Uses ranged from pastoral management of saltmarsh to the samphire (*Salicornia* spp.) gathering dealt with in Chapter 5. Even enclosure was probably not permanent and other uses such as hunting and fishing left no permanent marks on the landscape. In the absence of intensive use, these areas continued to maintain their nature conservation interest alongside the traditional management. The threat from malaria in what was often considered to be a mosquito-infested swamp occupying the margins of many estuaries or deltas, particularly in warmer climates, prompted drainage. Improved techniques of embankment and enclosure enabled more intensive uses to take place within these marginal areas, including arable production. As permanent enclosures were built, using the natural tendency for estuarine shores to accrete more saltmarsh was embanked ("primary land claim" Chapter 5, Doody 1995). Other wetlands were similarly affected with lasting changes such as the conversion of saline lagoons to rice fields or salinas ("secondary land claim" Chapter 11, Doody 1995) taking place.

This process of land claim was extended to encompass the winning of land from the sea through infilling with sediment and/or tipping of inert waste. The enclosure of open shallow seas has occurred extensively in Holland to create freshwater lakes and other non-tidal land. The combined effect of all of these activities has tended to restrict the area over which the processes of tidal movement, sediment erosion and deposition take place.

Exploitation of the living and non-living resources, particularly in estuaries, and their use for the disposal of a variety of waste materials, causes further pressure on the system and its wildlife. For the purposes of this chapter those wetlands which have developed on low-lying coasts where abundant sea-borne sediment collects in the wide mouths of tidal inlets and embayments or behind barrier islands are included, as are deltas which have developed in response to the accumulation of large amounts of material derived from land-based sources.

The coastal wetlands considered here are dynamic; they exist as a series of inter-related habitats and are frequently the locations favoured for human settlement, exploitation, and industrial development. The way in which these factors affect change, and the implications for their natural values and human use, are discussed in this chapter. The management of fjords, fjards and rias, more stable types developed in 'hard' rock landscapes, is considered briefly. Many key issues affecting these ecosystems have a marine bias and are not covered in this book.

10.1.1 Defining the ecosystems

Estuaries, deltas and lagoons are usually defined according to their physical structure, tidal regime, freshwater inflows and sediment characteristics. Depending on the definitions used, the general term 'coastal wetland' can include fjords and fjards which develop on largely glaciated rocky coasts, estuaries, including rias, wide-mouth tidal inlets, embayments and lagoons and deltas. Each of these is described in various text books; see for example estuaries (McLusky 1971, Dyer 1972, Day et al. 1998), deltas (Suter 1994) and lagoons (Barnes 1980). In the absence of human interference (though this is rare), individual wetlands develop in response to a complex set of interactions relating to physical location, relative sea level change, tidal range and pattern of inundation, sediment inputs/outputs and movement, and freshwater flows. Each of the main systems is defined below and an attempt is made to show the main features in Figure 10.76.

An Estuary is usually defined as:

"a semi-enclosed coastal body of water which has a free connection to the open sea and within which sea water is measurably diluted with fresh water derived from land drainage", Pritchard (1952) or

"an inlet of the sea reaching into a river valley as far as the upper limit of tidal rise", Fairbridge (1980).

Estuaries are most extensive in coastal areas with macro-meso tidal range and/or an abundant supply of sediment; in Europe these are especially significant around the southern North Sea. The relationships are complex and different types frequently overlap; the following definitions provide a useful distinction for management purposes:

1. **Drowned river valleys**. These develop where the land was inundated by the sea when sea levels rose rapidly towards the end of the last glaciation. The rapid advance of the sea which took place between 15,000 and 7,000 years ago (Warrick 1993) has been estimated to have been about 20mm per year. At its height some 8,000 years ago, in northwest Europe, when a catastrophic melting of the ice cap took place, it may have been as much as 75mm per year (Tooley 1993). The 'pulse' of sediment mobilised and moved landward by these events has helped create the extensive low-lying coastal plains where sediment movement plays an important part in the development of the estuary structure.
2. **Bar-built estuaries**. Similar to (1) above but where tidal influence and hence sediment penetration is restricted. This may be caused by the nature of the hinterland but usually stems from the presence of spits, bars or barrier islands which surround the river mouth and help create embayments. Because of the restricted opening to the sea, some of these may more closely resemble lagoons, see below.
3. **Rias**. Tidal inlets which are closely related to drowned river valleys though here the steep-sided nature of the land results in a much narrower profile. Depending on the elevation of the adjacent land, these estuaries may reach some distance inland with a series of side arms and are developed on rocky coasts.
4. **Fjords** and **fjards** which occur in glaciated, rocky coastal areas. Fjords with very steep walls occur in areas of high relief. Fjards occur in areas of low relief and may include shallow marine waters and small islands (skerries). For both the hard rock nature of the land results in a limited sediment supply.

Wetlands associated with barrier islands are not specifically identified here but form a landscape which can encompass both drowned river valleys and bar-built estuaries. They may also provide a link with deltas and lagoons and it is sometimes difficult to distinguish between the various types of system. They appear to occur mostly on meso-tidal coasts where wave energy is relatively high (Pethick 1984) and in micro-tidal zones.

Deltas are sedimentary coastal plains built up from material brought down by rivers to cover shallow offshore areas, creating sedimentary features which protrude beyond the normal limit of the coastal margin. They occur in at least six different forms depending on the relationship between river discharge and wave and tidal processes (Wright 1977). They are best developed in micro/meso-tidal areas where wave action is the dominant force in building the sand dunes, barrier islands bars and saltmarshes (Figure 10.76). They are most simply defined as follows:

"Marine deltas form where a fluvial system delivers sediment faster than marine processes can rework it." (Suter 1994).

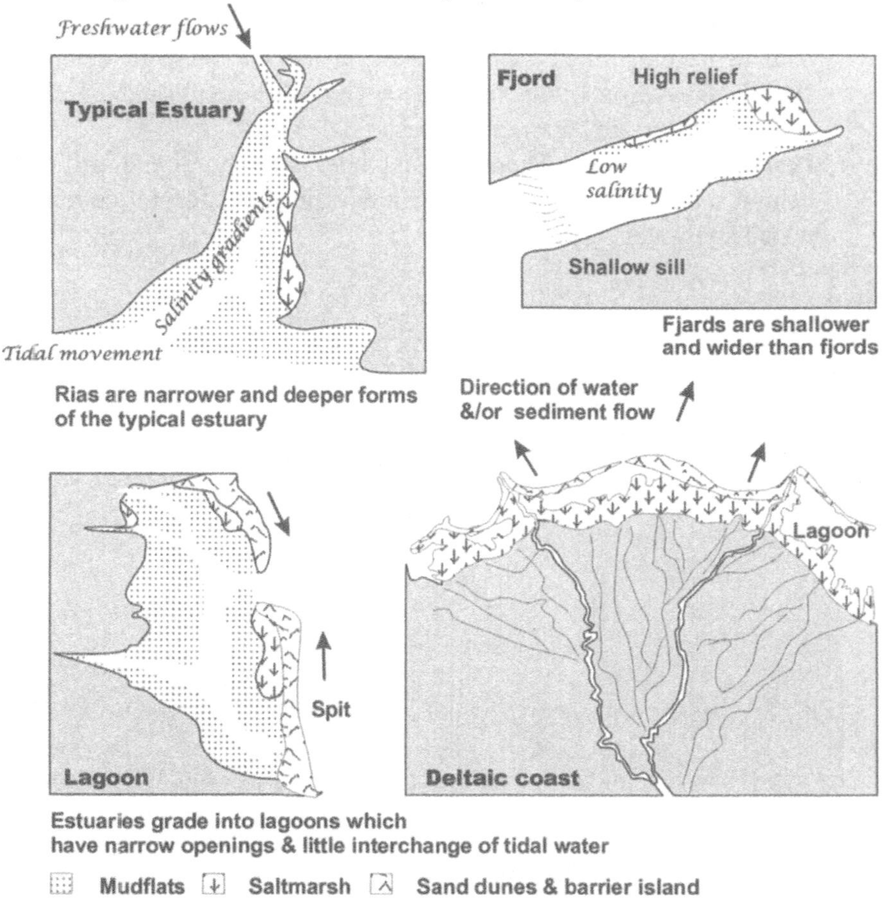

Figure 10.76. Some examples of coastal wetlands dealt with in this Chapter. No attempt is made to show all the types or gradations between them

Lagoons. The sometimes extensive lagoons, which may be associated with deltas and barrier island coasts, are a special form of wetland and defined as being:

"shallow enclosed or semi-enclosed bodies of sea water with a narrow entrance to an adjacent sea."

Lagoons are most common on micro-tidal coasts (Cooper 1994). Four main types are identified (Barnes 1980, 1996):

1. **Estuarine lagoons**. These merge into estuaries which have had their mouths partially blocked by offshore barriers, spits and bars;

2. **Bahira lagoons**. Pre-existing coastal embayments drowned by post-glacial inundation;
3. **Typical lagoons**. Lagoons developed behind barriers or spits cut off form the sea to form partially enclosed bodies of sea water. These include the Limans (lagoons and salt lakes) of the Black Sea coast. Although these tend to develop in micro-tidal areas in association with deltas (Figure 10.77), they also occur in areas where sufficient sand or shingle is deposited to create a barrier on meso-tidal coasts.
4. **Percolation lagoons**. These may include both natural and artificial areas behind sea walls or longshore barriers which cut off the tide but through which salt water seepage occurs.

Figure 10.77. The 'Fleet', a coastal lagoon, behind Chesil Beach, Dorset, England

10.2 Habitat distribution

Coastal wetlands are some of the most important wildlife habitats in the world. Reference to the comprehensive world-wide review of wetlands undertaken in association with the International Union for the Conservation of Nature (Dugan 1993) gives an indication of their extent. The maps associated with this publication show major coastal wetlands in northern Canada, east and southeast USA (including Chesapeake Bay and the Everglades) and in Europe the Wadden Sea and the delta area of the southern

Netherlands, the Nemas Delta in southeast Baltic, the Ebro, Rhone, Po and Nile Deltas in the Mediterranean and the Danube Delta in the Black Sea.

10.2.1 Wetlands in Europe

Coastal wetlands in Europe include example of all the types defined above. Fjords, fjards and rias occur on the glaciated exposed western shores of northern Europe in macro- or meso-tidal areas inundated at the end of the last glacial period. Major sedimentary systems occur in the southern North Sea where there is an abundance of sediment, much of it derived from soft material deposited during the sequence of glaciations of the Pleistocene. Deltas occur in the Mediterranean and Black Seas where sediment derived from erosion of the hinterland is deposited at the coast. Figure 10.78 shows the distribution of the major types of coastal wetlands. This is based on a variety of sources (e.g. Davidson et al. 1991, GB; Tooley & Jelgersma 1993, Europe and the Black Sea Environmental Programme (1997).

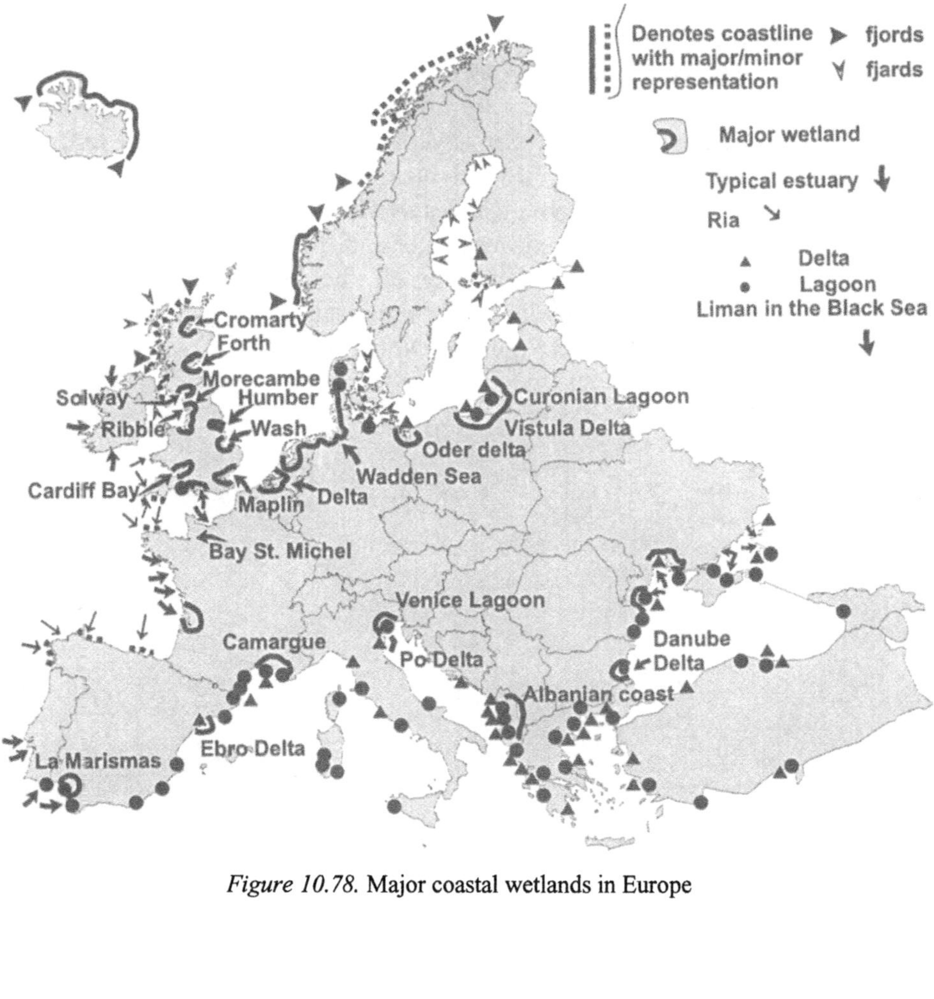

Figure 10.78. Major coastal wetlands in Europe

10.3 Nature conservation value

The nature conservation value of coastal wetland is derived from the complex interaction of its physical structure and function with biological and human influences. Wetlands are often highly productive and support a wide range of species commensurate with the range of habitats, as well as large populations of a few specialist species (e.g. marine invertebrates and wading birds).

A key component in the development of nature conservation interest in coastal wetlands, especially tidal estuaries, is their high productivity. This is derived from *in situ* primary production of saltmarshes and in estuarine waters, as well as inputs from the land and the sea. This abundant source of food can result in very large numbers of organisms. Typical densities on predominantly muddy shores for the mud snail (*Hydrobia ulvae*) and the burrowing amphipod (*Corophium volutator*), often numerically dominant, are up to 100,000 and 60,000 individuals per m^2 respectively (Barnes 1974). The segmented worms ragworm (*Nereis diversicolor*) and on more sandy shores lugworm (*Arenicola marina*), are also often abundant and of particular importance as prey for wintering wading birds.

Wintering bird populations are often the most obvious visible manifestation of the productivity of a wetland system. However a range of other species occur, including sea fish using the shelter of the estuary to spawn or as nursery areas. Some species of fish (diadromous) use the tidal waters as transitory area between freshwater and the open sea. There are also many resident species which may rely on the estuary for food whilst breeding inland. The main biological components of a typical estuary in northwest Europe are shown in Figure 10.79 below.

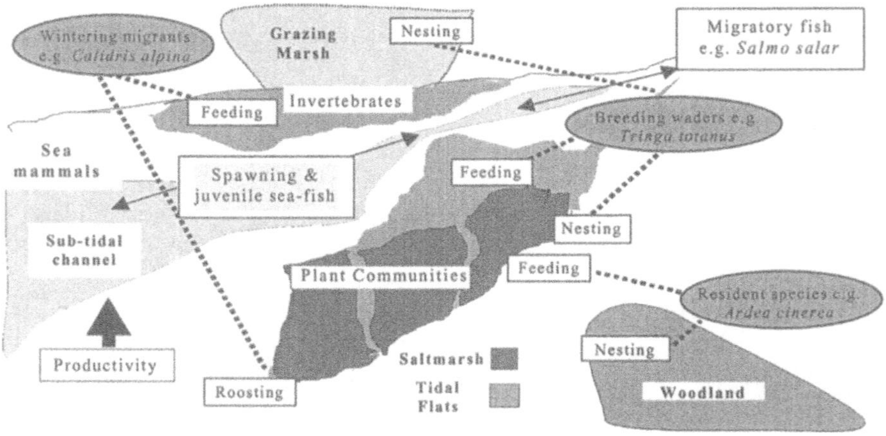

Figure 10.79. Principal components of nature conservation interest in a typical estuary

10.3.1 Plant communities

Plant communities of estuaries and deltas are mostly considered in the appropriate habitat chapters (Chapters 5 saltmarsh, 7 sand dune and 8 shingle). However, a number of taxa are of special conservation interest. *Zostera* is an important plant of the low shoreline especially on tidal muds. There are two species distributed across Europe and they provide important grazing for a variety of herbivores including notably, grazing ducks and geese. In addition to this they also contribute to the stability of tidal flats, a characteristic which they share with *Ruppia maritima*. To these should be added *Posidonia oceanica* and many other genera and species of sea-grasses, formerly terrestrial flowering plants now living as marine species. These have important value to coastal wetlands in the Mediterranean and elsewhere providing oxygen, food and shelter for a wide variety of marine species, nursery areas for commercial fish and helping to stabilise shorelines.

10.3.2 Invertebrates

The invertebrate fauna of most estuaries is usually poor in species, though those present are often abundant. Deltas and lagoons have similar faunas with species in common. Lagoon invertebrates and fish can be grouped into three elements according to Barnes (1996):

1. freshwater species able to withstand various degrees of brackishness;
2. marine/estuarine species able to tolerate low salinities;
3. specialist lagoonal species restricted to this habitat and some inland seas and tideless brackish ponds and ditches (Table 10.27).

Table 10.27. Some characteristic invertebrates of lagoonal habitats in northwest Europe

Key indicator (i.e. Specialist) Species, * Extinct?	
Hydroids	**Prawns**
Clavopsella navis	*Palaemonetes varians*
Sea-anemones	**Gastropod Molluscs**
*Edwardsia ivelli**	*Hydrobia ventrosa*
Nematostella vectensis	*Hydrobia neglecta*
Polychaetes	*Hydrobia acuta*
Armandia cirrhosa	*Hydrobia minorcensis*
Alkmaria romijni	*Heleobia stagnorum*
Isopod Crustaceans	*Littorina saxatilis* var. *lagunae*
Idotea chelipes	*Tenellia adspersa*
Amphipod Crustaceans	**Bivalve Molluscs**
Gammarus chevreuxi	*Cerastoderma glaucum*
Gammarus insensibilis	**Bryozoans**
Corophium insidiosum	*Victorella pavida*
	Conopeum seurati

10.3.3 Fish

A wide variety of fish use coastal wetlands, including many of commercial importance. They range from some pelagic fish (living in or near the surface of the open sea) which spawn in nearshore coastal waters and use estuaries and deltas as nursery areas, to species which simply pass through on migration. The northwest sea lochs (fjords and fjards) provide nursery areas for juvenile herring (*Clupea harengus*) and sprat (*Sprattus sprattus*). In the southern North Sea herring spawn early in the year inshore in the Thames estuary where they attach their eggs to coarse-grained material in the outer reaches of the estuary. Many demersal fish (living on or near the bottom of the sea) are also reliant on coastal wetlands; plaice (*Pleuronectes platessa*) and several mullet species are closely associated with estuaries and deltas. Sea bass is an important species which relies on sheltered estuaries and deltas as nursery areas for juvenile fish. These coastal wetlands often have extensive and intensive aquaculture, especially in the Mediterranean with common species such as the common eel (*Anguilla anguilla*) and grey mullets (*Liza ramada* and *Chelon labrosus*), sea bass (*Dicentrarchus labrax*) and sea bream (*Sparus aurata*).

Diadromous fish are specialists of the estuarine environment. They include catadromous species, such as the common eel, move from freshwater to the sea where they spawn, passing on the way the salinity gradients in the estuaries and deltas. In contrast, an andromous species such as salmon (*Salmo salar*) develops from eggs shed upstream in rivers. A major part of its life is spent outside the home rivers at sea. Important feeding areas are in the Atlantic along the west coast of Greenland, above the Faroe Islands, in the Norwegian Sea and possibly in the Barents Sea. In the Baltic the Curonian Lagoon has 46 freshwater, migratory and marine fish including the Baltic salmon, reflecting its importance for these species (Repečka 1994).

10.4 Species distribution within an estuary

The range and complexity of the coastal wetland environments require special consideration when devising management strategies. In addition to the spatial and temporal relationships between habitats, many species range widely both within and across the ecosystem. It is important to look at the full complement of biological diversity in order to ensure that management in one area does not compromise interests in another. In this context it is instructive to consider the way in which individual species range across the tidal zone. Figure 10.80 attempts to depict, for a few representative estuarine species, the habitats on which they depend. This schematic representation takes no account of the further spatial relationships involving migratory species, dealt with in chapter 12.

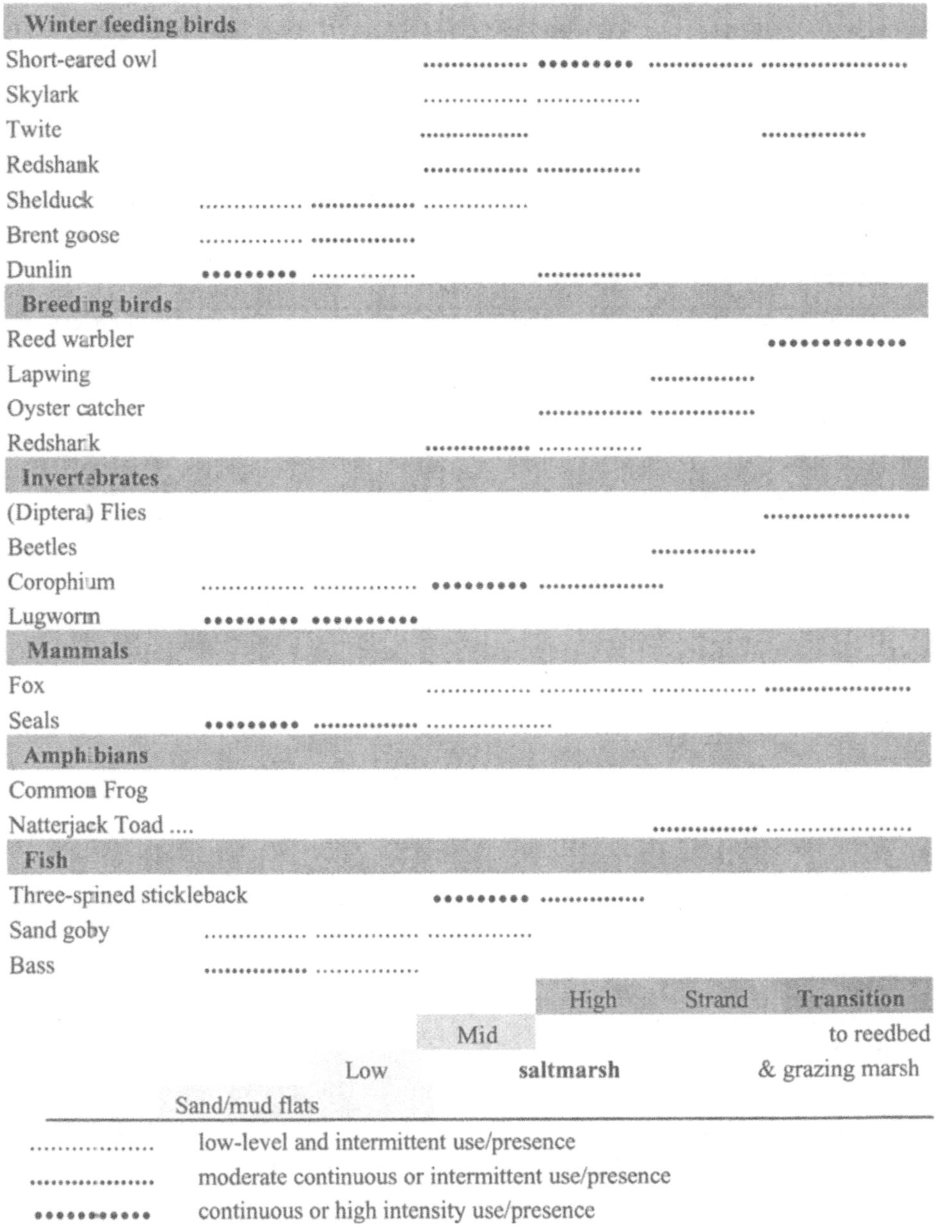

Figure 10.30. Distribution of some key animal species or groups of species across a typical estuary transition where there has been no enclosure. NB This figure does not take into account all locations where species may occur in response to tidal movement

10.5 Human activities and conservation

As has already been stated, estuaries, deltas and lagoons were amongst the earliest coastal areas to be settled (Chapter 1). This occupation and development in the hinterland have had a major influence on them ever since. Drainage and sometimes large scale enclosure has frequently reduced what was once a wide flexible zone, able to respond to changes in sea level rise (in the long term) and storms (in the short term), to a narrow fringe often bordered by artificial defences. These and other human activities are listed in Table 10.28 below. The way these gross changes have influenced the conservation of coastal wetlands is reviewed below.

Table 10.28. Key management issues affecting coastal wetlands, taken mainly from examples in northwest Europe

LOSS RESULTING FROM:	**REMEDIAL ENGINEERING:**
Industrial, port & harbour development;	Sea walls, earth banks & groynes;
The creation of grazing marsh (Chapter 11);	Other protection of barrier island, dune and
Creation of salinas (Chapter 11);	saltmarsh;
Saltmarsh enclosure (Chapter 5);	Excavation of upper saltmarsh;
Drainage of swamps and fens;	Polders.
Intensive agricultural use;	**ACCESS FOR SPORT AND RECREATION:**
Conversion of salinas to rice fields.	Walking & bird-watching;
HYDROLOGY:	Wildfowling;
Saltwater intrusion & pollution	Boating.
Water abstraction & dams;	**POLLUTION:**
Irrigation.	Oil & sewage;
DREDGING, EXCAVATION &	Chemical & from the air;
SEA-LEVEL RISE:	Litter.
Erosion & foreshore steepening;	**OTHER USES:**
Effects on vegetation succession;	*Spartina* planting/management (Chapter 6);
Breeding birds;	Fisheries (oyster culture in saltmarshes);
The saltmarsh 'squeeze' (Chapter 5).	Fish farms, aquaculture (especially deltas).

10.5.1 Historical loss

The loss of coastal areas has already been discussed in general terms in (Chapter 1) and more specifically in relation to saltmarshes (Chapter 5). Major enclosures in the Wadden Sea and southwest Holland have occurred this century, and up until the last decade enclosure for intensive agricultural use in the Ribble Estuary (Figure 1.3) shows the continuing and cumulative nature of the developments. Even quite small losses can add up to major change. Drainage of marginal swamps prompted by the need to reduce the prevalence of malaria and subsequent cultivation has reduced the area of many coastal wetlands throughout Europe. The Venice Lagoon settlement began about 450 AD and since then some 2,500ha has been lost to agriculture, 2,700ha to industry and 8,200ha converted for fish-farming, out of a total area of some 58,000ha (Macdougall 1996). Even the coastal lagoons and wetlands of Albania (one of the least affected coastlines of any Mediterranean country) have lost 15,000ha, mostly to agricultural use in the last 40 years - more than 20% of the total 70,000ha according to Gjiknuri & Hoda (1996). Today many deltas have been reduced to a narrow fringe of uncultivated land through drainage and conversion to agriculture. Lagoons are used for commercial salt production and/or converted to rice fields (Chapter 11).

On the more industrialised coastlines tipping of inert material, including refuse disposal, or the pumping of intertidal sediments into the void behind a newly erected sea wall on tidal flats, has provided consolidated land suitable for building or other development. Industrial uses include the expansion of port facilities, roads and housing. Cardiff Bay, south Wales (Figure 1.4) was finally completely destroyed as a tidal estuary with the closure of a barrage to create an amenity lake. In the USA urbanisation is the principal cause of the loss of tidal land around cities such as Boston and New York.

10.5.2 An accumulation of losses

The enclosure of saltmarsh for agriculture and the loss of tidal land to industry is often piecemeal but may involve large individual works dictated by particular economic pressures. In the UK water resource needs resulted in consideration of several proposals to store freshwater in impounded reservoirs situated in large embayments including the Wash and Morecambe Bay. The need for a third London Airport prompted detailed environmental studies of the implications for Maplin Sands, an important wildlife area in southeast England (Macey 1975). None of these proposals has gone ahead because of economic and environmental considerations, but cumulative losses from other activities have caused wide-spread losses to many other sites (Figure 10.81).

Other developments include barrages such as the amenity barrage described for Cardiff Bay (Chapter 1, Figure 1.4). These destroy the tidal land completely. Tidal energy or storm surge barriers are less destructive as some elements of tidal movement remain. The latter have differing impacts depending on the time over which they operate and the responses of the tidal and sedimentary regime to their operation (Gray 1992).

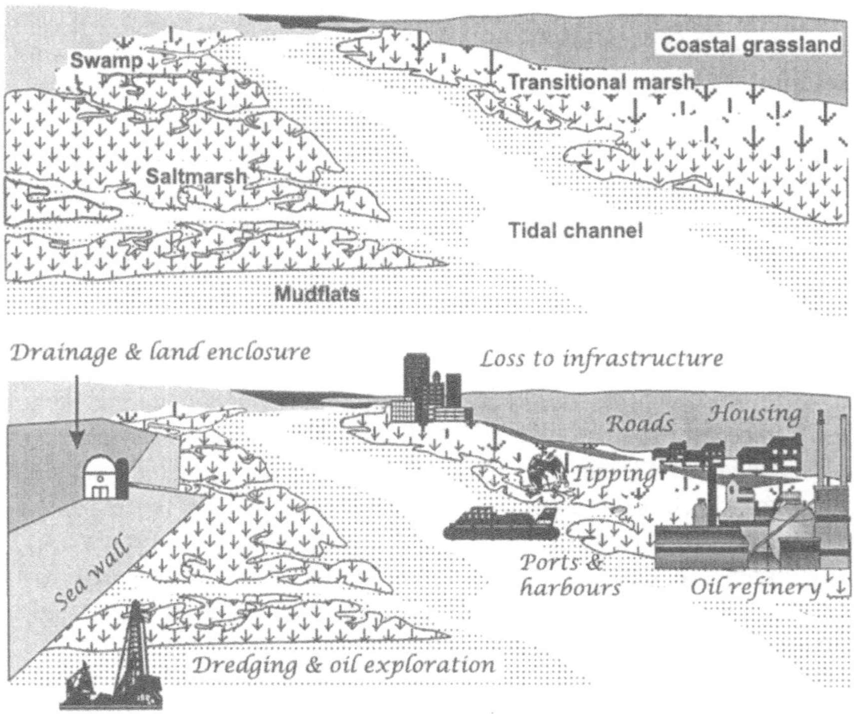

Figure 10.81. Loss of estuarine habitats, typical situation in Europe

The threat from agricultural development has receded in many areas as a result of changes in economic factors, conservation legislation and the recognition of the implications of sea level rise. In the Wash proposals for new enclosures were tested by Public Inquiry in the early 1970s. A resulting 10 year moratorium was superseded by County Structure Plans which included policies with a presumption against enclosure. Although the conservationists lost the economic case, circumstances and government policies towards agriculture changed such that the pressure for enclosure lessened. No enclosures have taken place since. In the German Wadden Sea a similar sequence involving economic change, legislation and greater environmental awareness, including the implications of sea level rise, has also led to a gradual decline in wetland loss (Goeldner 1999). However, whilst these developments have receded, others have taken their place. Amongst the most significant of these in the United Kingdom are proposals for a tidal power barrage across the River Severn in southwest England and south Wales which would have completely altered the tidal environment of the estuary. Most recent of all is a proposal to extend the airport of Schipol

(The Netherlands) into the sea, though this too has been dropped for the time being.

Coastal wetlands are, and will continue to be, the focus for major economic developments. There is no sign that these pressures will be reduced and the consequences of further developments are unpredictable. Extrapolations for expansion of coastal cities, many of which are located on or near estuaries or deltas (Hinrichsen 1998), suggest that major losses of land will continue to occur in the future, adding to the economic and environmental problems for these areas.

10.6 The 'estuary squeeze'

Habitat loss is only one of a number of factors which puts pressure on the conservation value of estuaries and other coastal wetlands. As these areas diminish through land-claim the seaward margin may be affected in several ways which can exacerbate the effects of the loss of land. These pressures take several different forms, ranging from changes in sea level and sediment availability to the nature of human exploitation. Where high quality agricultural land and/or infrastructure are threatened the response to potential erosion and flooding is critical to the way the coast itself reacts. There will be a natural desire to protect life and property from the impact of the sea. However, maintaining the precise line of defence set out in earlier times may not be in the best interests either of the coastal environment or of its ability to sustain human use. Two principal issues deserve further consideration, sea level change and sediment availability.

10.6.1 Sea level change

One of the major determinants of the nature of the coast lies in the rate and direction of sea level change. From about 10,000 years ago at the end of the last glaciation rapid changes in sea level occurred and had a major impact on the nature of the coastline. When sea level began to stabilise, at about its present level about 7,000 years ago, the coastal formations that are present today began to develop. Their evolution, then and now, depends on such factors as the availability of sediment, the influence of sea level change, tidal range and wave action. Overall there is a general trend towards erosion and mobilisation of sediments, as global sea levels are thought to be rising at about 1-2mm per year (Warrick & Oerlemans 1990). Exceptions occur in areas where isostatic rebound of glaciated areas is greater than the present rise in sea level, causing a general elevation of the land relative to the sea in areas such as Norway, the Gulf of Bothnia and around the margins of the

Central Highlands of Scotland (Carter 1988) or in areas where seismic activity raises the land.

The way in which coastal wetlands respond to changing sea levels is an important consideration in their conservation. Typically tidal estuaries and deltas are amongst the most responsive ecosystems. Low-lying coastal plains with abundant supplies of sediment derived from sea and/or land move landward or seaward depending on the direction of relative sea level change. Rising sea levels result in landward migration of the shoreline, falling sea levels an extension of the shoreline as new habitat accretes to seaward. These effects are of course modified by the availability of sediment which, when particularly abundant, can force a seaward movement even if sea level is rising. The cumulative effects of sea level rise and subsidence on coastal lowlands have been considered at a global scale (Bird 1993), for Europe (Tooley & Jelgersma 1993) and for individual sites such as Venice, and the Niger and Mississippi deltas (Milliman 1996).

10.6.2 A question of sediments

The availability of sediment is of critical importance to the long-term viability of many coastal wetlands. Material may be derived from erosion in the hinterland (often resulting from deforestation), undercutting and collapse of soft coastal cliffs and from glacial and other soft sub-sea deposits. Reduced sediment supply (caused by offshore extraction, protection of eroding cliffs and damming of sediment-rich rivers) has contributed to the further loss and degradation of coastal wetlands, particularly at the seaward margins, throughout Europe. This is most pronounced in the deltas of the Mediterranean. The Ebro Delta, for example, has grown through the transport and deposition of sediments eroded from the hinterland following deforestation. Studies suggest that at the end of the last century this situation began to be reversed as the damming of rivers reduced the available sediment supply. Today the outer margins of the delta are eroding as it becomes a wave-dominated shoreline rather than one where fresh-water river flows exert the major force (Palanques & Guillén 1998). The building of the Aswan Dams has effectively cut off the sediment supply to the Nile Delta and the shoreline is retreating as a result (Stanley & Warne 1993). These examples continue an historical pattern which has affected many, if not all, the deltas in Italy. In Venice urgent consideration is being given to using dredged material in combating rising sea levels and reversing the net export of sediment from the system (Day et al. 1998).

Although there are many examples of retreating shorelines within coastal wetlands, there are still a few countries where abundant sediment reaches the sea. The deltas of Albania and southeast Turkey, for example, continue to grow because of deforestation in the hinterland. To date there are few dams to stop rivers delivering sediment to the coast, though this is changing. In some areas land enclosure does not always result in an overall loss of tidal habitat, even where relative sea level is rising. In estuaries such as that of the River Dee in northwest England there is an overall trend towards siltation (Pye 1996). This seems to be related to the reduction in tidal volume and scour which allows sediments to be deposited within the estuary. A similar situation occurs in the Bay of St Michel in western France where the tradition of enclosure seems to have stimulated the movement of tidal flats seawards.

Whatever the mechanism occurring in each estuary or delta, as enclosure takes place the extent of the flexible margin between the land and the sea is diminished. The main factors contributing to a loss of intertidal land on the estuary foreshore is illustrated in Figure 10.82. In those areas where sea level is rising relative to the land eroding shores become steeper and the problems are exacerbated. In many areas saltmarshes are the first to suffer as erosion and slumping take place. Even where mudflats and saltmarshes are accreting it appears there may be a limit to the position of low-water mark (Chapter 5). Overall the effect is to diminish the ability of the shoreline to accommodate perturbations due to factors associated with global warming, principally involving long-term sea level change and an increase in the frequency of storms.

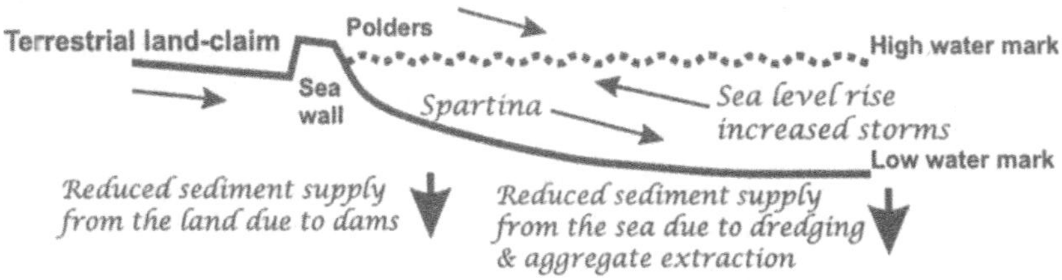

Figure 10.82. Main factors affecting an 'estuary foreshore squeeze'

10.7 Coastal defence

Against this background it is relevant to question the usual response to threats of erosion and flooding to high quality agricultural land and major infrastructure developments. Despite the increasing cost of sea defence or coast protection (particularly in areas of rising sea levels), sea walls and other artificial coastal protection features continue to be built or 'improved' in order to maintain the current line of defence. Where estuaries or deltas are sinking, then the structures themselves become less and less sustainable and more costly to repair.

10.7.1 Polders

Encouraging the growth of saltmarsh by the development of sediment fields has been a major activity, extensively used in Denmark, Germany and Holland (Kamps 1962). Faced with loss of saltmarsh and the undermining of the adjacent sea walls in southeast England, attempts have been made to recreate them there also. The protection, or recreation of saltmarsh, is on the face of it a good thing for both the protection of the sea wall and the restoration of natural habitat. However, balanced against this must be the foreshortening of the intertidal profile which cannot retreat landwards because of the existence of the sea wall (Pethick 1996). Promoting the growth of saltmarsh may in its turn have knock on effects for the availability of tidal feeding area for wintering waterfowl (see the discussion on *Spartina* in Chapter 6). Before further work is undertaken a thorough review of the implications of these techniques for nature conservation should be made.

10.7.2 Sediment extraction and dredging

In some areas extraction of sand and gravel takes place outside the confines of an estuary, often offshore. 11 out of 155 estuaries in the "Estuaries Review" (Davidson et al. 1991) involve the extraction of commercial quantities. Figures for northwest Europe give some idea of the scale of this activity. In 1994 the amount of aggregate extracted from offshore sources was estimated to be 20 million tons (England and Wales), 13 million tons (Holland), 5 million tons (Denmark) and 2 million tons (Belgium) (Harrison 1996). It is not clear, given that a large proportion is devoted to beach nourishment, what the long-term consequences of this level of activity are for coastal stability. There will be some environmental impact, not least by accelerating natural changes in sediment regimes. This may be particularly problematic on coastlines which are eroding, though not exclusively.

At a few sites in the United Kingdom (e.g. the Ribble Estuary, Lancashire) sand is extracted at the mouth of the estuary. Changes in the sedimentary patterns in the Ribble have affected the nature of the beach sediment on the nearby Sefton coast. The sandy beach, much used for recreation, has and is changing into one with more silt and a muddy texture. Invasion of *Spartina anglica* has helped lead to a perception that there is a causal link between sand extraction and the changing beach sediment. This link has not been established and changes to the tidal volume within the estuary consequent upon enclosure may also be a factor. The impact of channel dredging, a more widespread activity, is similarly not well understood. Both of these activities would repay greater investigation in relation to their effects on the sediment regime within an estuary and the knock-on effects on wildlife.

10.8 Pollution

Coastal wetlands, including estuaries, have been used for many years as a convenient place to dispose of sewage and other pollutants. The enriching effects of some of these effluents, when combined with the run-off of agricultural fertilisers, can cause excessive growth of green algae. This may help to decrease the diversity of life in the estuary and even have a detrimental effect on saltmarshes, as upper vegetation is smothered by the excess plant material.

Sewage inputs increase the demand for oxygen and result in an overall decrease in species diversity in the water column and tidal flats (McLusky 1971). Other, toxic pollutants may kill fish and in some instances can cause waterfowl mortality. In the Mersey Estuary in northwest England, inputs of organic matter have resulted in profound changes with a loss of invertebrates due to anoxic conditions. This, in its turn, is thought to have been the major reason for a decline in fisheries in the 1960s (Hawkins et al. 1999). More acute poisoning has been observed in waterfowl with deaths occurring due to lead accumulation (Bull et al. 1983).

In the Baltic, lagoons are used for rearing fish, such as grey mullet and other species. Losses due to pollution from 1981-1993 included greatly reduced numbers of migratory fish such as twaite shad (*Alosa fallax*) vimba (*Vimba vimba*) and salmon; such fish were also affected by damming of rivers for hydro-electric power. (Repečka 1994).

10.9 Time-scales for change

Coastal wetlands are complex and dynamic systems where human use is frequently an integral part of their development. The way in which these complexities can be managed in an integrated approach is dealt with in Chapter 13. Here it is valuable to consider some overarching principles pertaining to the wetlands themselves. Amongst these understanding the timescales over which coastal processes operate will be critical when attempts are made to manipulate whole systems.

Coastal wetlands develop over timescales measured in thousand years. Change at the margins may occur over periods of only a few hundred years and the processes which bring about these changes operate over much shorter time scales. Tidal cycles change over periods of 10 days and beach-ridge formation, barrier migration and cliff erosion between 10 - 100 years (Carter 1988). Estuarine mud can be transported 16km in less than 24 hours whilst heavier material such as pebbles on a shingle beach can be moved more than 2km in less than a month (Steers 1981).

10.9.1 Successions and transitions

Recognising the changing nature of the estuary, and the time scales over which the various underlying processes operate, may be just the first step towards a more enlightened approach to management. Natural forces, such as sea level change and storms, operate over time-scales from several hundred years to only a matter of hours or minutes respectively. Many individual plant and animal species are adapted to daily changes in tidal movement, salinity and temperature. Sediment deposition influences the nature and rate of succession between and within habitats. In tidal estuaries, for example, a mudflat can be colonised by *Spartina anglica* and in the absence of grazing become dominated by *Phragmites australis* in only 10 years (Ranwell 1972a). Such a rapid succession is not the norm; and upper estuary or high level saltmarsh, and transitions to brackish and fresh marsh, amongst the most diverse of the tidally influenced habitats, attain maturity over periods up to 330 years or more (Chapman 1974).

Many of the animal components within the estuary are specifically adapted to, or tolerant of, change. These communities characteristically have a low species diversity. Individual species have high fecundity and are thus able to colonise areas rapidly following perturbations in environmental conditions. For example re-colonisation quickly took place following a two hundred year storm event in Chesapeake Bay in 1972. This single event brought sediment into the Bay equivalent to inputs for the previous decade and the high freshwater flows persisted during the 4-5 days of the storm. Despite this major change in the ecosystem, plankton communities recovered within 2-3 weeks, benthic fauna within a year and there was no apparent effect on the annual fisheries yields (Costanza et al. 1993).

10.9.2 The margin of error

It is against this background of both short and long term changes to the nature of the coastal wetland shoreline that management and exploitation must be considered. Today the ability of the coastline to provide a flexible response to changing patterns of sea level rise and storms, is prevented in many areas by the creation of a straight inflexible barrier at the margin of the estuary. In a zone which is recognised as being subject to the forces of nature, it could be argued that man has devoted too much effort to his 'battle with the sea', squeezing the coastline into an ever-narrowing zone. This could have grave consequences both for the survival of the wildlife which inhabit these areas and for the future sustainability of human use.

Recent studies have also shown the importance of episodic events to the delivery of sediments to these systems, particularly deltas (Day et al. 1997). Their models show the significance of 'pulses' of sediment in combating subsidence and sea level rise. They argue that the restriction of sediment supply brought about by the damming of rivers will, amongst other things, compromise both the economic and biological sustainability of these areas. This suggests that the margin of error for both is small. In the Ebro Delta in Spain, the relationship between sea level rise and sediment deposition in the delta, points to the need for partial removal of sediment collected in dams to raise the elevation of the delta by 50cm (Ibàñez et al. 1997).

10.9.3 Levels of complexity

Coastal wetlands, especially estuaries, deltas, and lagoons, are complex systems in their own right and often exist in combination with other habitats. This chapter has been principally concerned with wider ecosystems and the way human use has influenced their ability to sustain wildlife. The underlying complexity of estuaries is described using 'generalised' food webs (Barnes 1974), 'major' nutrient pathways for individual habitats (e.g. sand dunes, Packham & Willis 1997 pp. 215-220). The inter-relations between these and the component species depend on further temporal change, as in the population dynamics of individual species and predator-prey relationships between them. When added to the spatial development of plant communities through succession and the migratory patterns of mobile species the number of potential interactions is considerable.

Despite frequent and major losses of habitat and interference with boundary processes and sediment cycles, most coastal wetlands continue to sustain wildlife and human use; a testimony to their resilience. However, as these wetlands are 'squeezed' spatially and more pressure is brought to bear on them, their ability to adapt to both natural and human-induced changes is impaired. Where the wetland is totally destroyed, as in Cardiff Bay (Figure 1.4 above), the results are painfully obvious. Replacement habitats may mitigate against the habitat loss but it is still impossible to predict how these complex systems will react to further losses. This is particularly important in assessing their likely response to changing patterns of tides, storms and sea level change consequent upon global warming. It may be that recent attempts to recreate coastal wetlands in Europe and North America may provide a more certain way forward. Wider coastal zones are in the best interest of wildlife and human use as they accommodate perturbations more efficiently than narrow ones.

11. GRAZING MARSHES & SALINAS

11.1 Introduction and scope

Few, if any, coastal habitats are truly natural, but the two principal habitats in this chapter have a special conservation significance because they have largely been created through human intervention. In nature conservation terms the habitats lie between those which are (or are considered to be) entirely natural and those which are intensively used, with little or no conservation interest. The former include inaccessible sea cliffs and newly formed sand dunes, the latter intensively cultivated grasslands and cereal crops. These 'secondary' wildlife habitats (as defined in Chapter 5) are dealt with separately below and their key attributes discussed.

11.2 Grazing marsh (coastal wet grassland)

Grazing marsh can include both wetlands subject to a maritime influence and lowland wet grassland adjacent to the margins of the coast, especially tidal reaches of estuaries. Grazing marsh is a special kind of coastal wet grassland. It is usually derived from the enclosure of saltmarsh and includes a number of ditches that drain either brackish or freshwater. The land so created was traditionally used for grazing, allowing pastureland to develop. A wide variety of plants and animals also co-existed within the relatively low intensity management. As techniques of enclosure improved, larger areas were more permanently excluded from tidal inundation resulting in increased use by domestic stock. Eventually as the salt leached out of the soil it became possible to use the land for arable crops, thus completing a process of destruction, including any residual nature conservation interests, that may have survived the original enclosure (see Chapter 5).

Today, despite drainage and losses due to ploughing, a number of important areas of coastal grazing marsh remain under traditional management. These are known mostly from the UK and continue to support a wealth of wildlife. Taken together with other wet para-maritime coastal grazing pasture, they form a valuable habitat which is highly susceptible to the impact of intensive agriculture. The first part of the chapter deals with those coastal grasslands derived from estuarine saltmarshes which are important for breeding and wintering waterfowl and/or have a significant invertebrate fauna associated with brackish water ditches.

11.2.1 Habitat definition

The partial enclosure of saltmarsh by the erection of a small earthen bank was probably one of the first human activities directly affecting coastal habitats. The aim was to extend the period of Summer grazing by livestock (cattle and sheep), by preventing over-topping on Spring tides. As the early techniques of enclosure improved, larger clay or earth banks were constructed enclosing consolidated saltmarsh excluding the tide from larger areas. Historically the land was subsequently used as pasture, with little or no further 'improvement'. These areas of coastal grazing marsh are defined by the presence of permanent and semi-permanent grassland, drainage ditches and enclosing earth dykes. The wildlife interest has developed alongside the traditional use of the area for agriculture. The principal components which help to define these systems include:

- Sea walls;
- Saline seepage areas;
- Vehicle rutted ground;
- Grassland (alongside the sea wall);
- Counter ditches (alongside sea wall);
- Grazing marsh;
- Cattle poached areas;
- Dredged spoil heaps;
- Fleet (former tidal channel);
- Reed beds and scrub (Gray 1977, p. 257).

The method of enclosure and origin of the pasture help to define the habitat type. Any or all of the components described above may be present on an individual site, depending on the management of the land. The more intensive the use the fewer components may survive. Hence the enclosed area will have agricultural land (arable crops may be present at one end of the spectrum, permanent unimproved pasture at the other), drainage ditches (highly eutrophicated or not; brackish to freshwater) and a sea wall (Figure 11.83).

Figure 11.83. Typical coastal 'grazing marsh', north Kent, England

11.3 Habitat distribution

The origins of these habitats determine their location along the margins of estuaries. Little is known about the distribution of coastal grazing marsh other than in the UK. No systematic surveys are available and grazing marsh is not recognised as a distinct habitat type in the European Union Habitats and Species Directive, or the Corine classification from which they are derived.

11.3.1 Habitat location

Substantial areas of tidal land have been 'won from the sea' through the process of saltmarsh enclosure described in Chapter 5. An indication of their potential extent can be made by reference to Figure 11.84a which shows the distribution of alluvial soils in Great Britain (redrawn from Boorman et al. 1989). Coastal wet grassland occurs extensively around Great Britain (Figure 11.84b, derived from Barne et al. 1995/8). Comparison of these two figures suggests that coastal grazing marshes, as defined above, have a much more restricted distribution. They are present around the margins of the estuaries of the east and south coasts of England and in south Wales. They are absent from most of Scotland and north Wales where alluvial soils are restricted. In east Anglia, especially around the estuaries of the Humber and the Wash where considerable land drainage and enclosure has taken place, they have long since mostly converted to intensive agriculture.

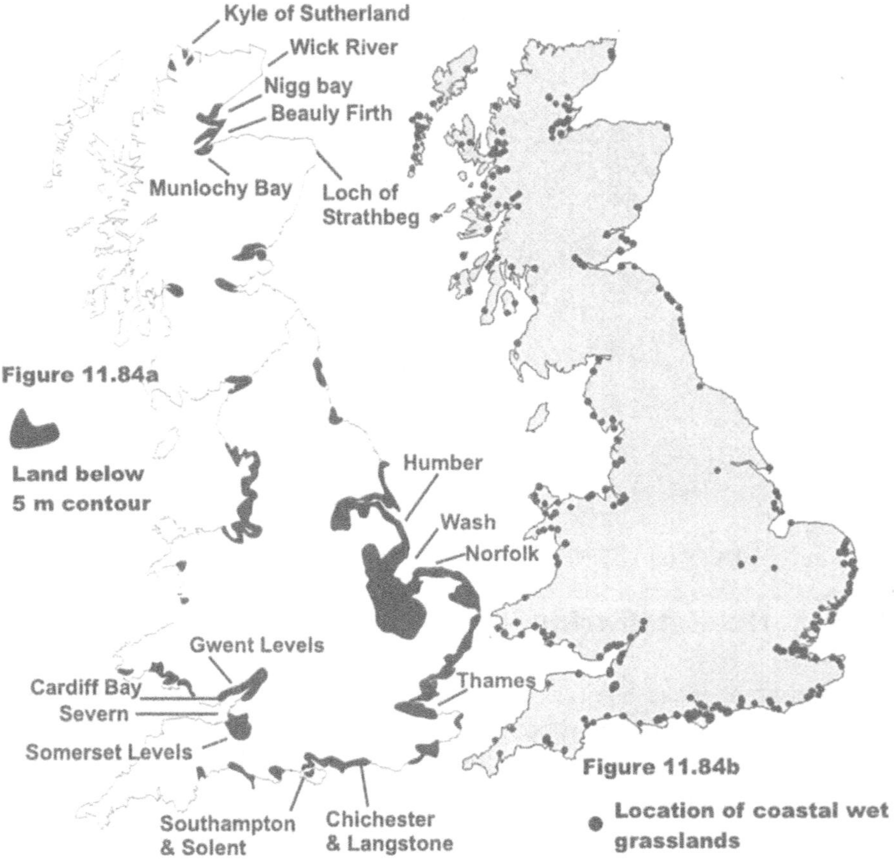

Figure 11.84. Coastal grazing marsh on alluvial soils in England and Wales in relation to unimproved or partially improved wet grasslands in the UK. For a and b, see text on page 211

The Thames Estuary, Southampton Water and Solent on the coast of Hampshire and the land surrounding the Severn Estuary (Allen 1992) are the main surviving locations. The last site includes the internationally important Somerset Levels and similar habitats on the Welsh side of the estuary. Other smaller sites do occur in Scotland and Northern Ireland though most are not derived from alluvial soils occurring as undrained coastal wet grassland.

11.4 Nature conservation value

The nature conservation interest includes common and rare coastal plant species, especially those which are typically found at the upper end of the tidal cycle on high level saltmarshes. Plant communities in the ditches range from those typical of brackish waters to those with little or no special coastal interest. Breeding and over-wintering birds, together with the flora and invertebrate fauna associated with the brackish to freshwater ditches (dykes), make grazing marshes a habitat of considerable nature conservation importance.

11.4.1 Plant communities

The plant communities of the drier areas are most frequently closely allied with neutral grasslands, but they include other important, coastal elements. Gray (1977) lists a total of 23 species which are common or rare and for which grazing marsh is the only or a major habitat. These include *Bupleurum tenuissimum*, a locally distributed species of dry grasslands and sea walls. A variety of other species of high-level saltmarsh such as *Puccinellia distans* and *Parapholis strigosa* are also distinctive.

In other areas, such as the wet grasslands surrounding the estuaries of the Solent on the south coast of England, a transition from base-rich fresh marsh to land-claimed saltmarsh occurs. At other sites the distribution of plant species may follow the changing underlying soils, from dry acid sward on the edge of the flood plain, dominated by *Festuca rubra*, *Holcus lanatus* and *Anthoxanthum odoratum*, to damper grassland dominated by *Agrostis stolonifera* and *Deschampsia cespitosa* and including *Menyanthes trifoliata*, *Baldellia ranunculoides* and *Butomus umbellatus* as on the Somerset Levels. The range of salinities in the ditches also supports a wide range of species. These include swamp species such as *Scirpus maritimus* and *S. tabernaemontani*.

11.4.2 Rare plants

One small area of coastal grazing marsh in Essex is the only locality in Great Britain for *Atriplex pedunculata*, a species which on the continent is scattered along the coast of Denmark to Northern France on more sandy soils in saltmarshes. Ditches in coastal grazing marshes are especially rich. For example on the north Norfolk coast they are a stronghold for the nationally scarce *Ceratophyllum submersum* and *Potamogeton obtusifolius* (Reid et al. 1988). On the south coast transitions from base-rich fresh marsh to land-claimed saltmarsh support the uncommon *Oenanthe pimpinelloides*. Further north in Scotland wet grasslands (mostly not derived from saltmarsh) along the Wick River support nationally important sedge communities, including the rare *Carex recta* (estuarine sedge also called the Wick sedge). It is known from only six locations in the British Isles, all in the Highland

Region and including the Kyle of Sutherland, where the ditches support a range of fen and wet grassland communities (Perring & Farrell 1983).

11.4.3 Invertebrates

Invertebrates are an important element in the interest of these areas with special and rare species often associated with the salinity gradients. Brackish water ditches and borrow-dykes behind sea walls are of particular value. Species-poor stands of *Bolboschoenus* (*Scirpus*) *maritimus* or *Phragmites australis* can support rich and important invertebrate communities. Earth sea banks, although poor replacements for the transitional habitats associated with upper saltmarsh, also have special interest. *Artemisia maritima* for example supports the specialist fly, *Paroxyna absinthii* and a jumping louse *Craspedolepta pilosa* (Kirby 1992). The grazing marshes of Essex and North Kent as well the Somerset and Gwent Levels are of national importance for their assemblage of invertebrates, including many rare and notable Coleoptera and Diptera.

11.4.4 Birds

In many areas of the UK, grazing marsh (and coastal wet grassland generally) is recognised as being of particular value for birds, notably for breeding and wintering waders and wildfowl (Davidson 1991). The Somerset Levels in southwest England are amongst the most valuable areas, and species characteristic of this wet grassland include the ducks: wigeon, teal and pintail (*Anas acuta*) and among the waders lapwing, redshank and snipe. Coastal grazing marsh around Chichester and Langstone Harbours in southern England support internationally important wintering populations of brent goose (*Branta bernicla bernicla*) and many other species, together with important breeding populations of several rare bird species of national interest.

In Scotland, coastal wet grassland is known to support good numbers of breeding waders on the Kyle of Sutherland and probably does so elsewhere. The Carse of Bayfield on the north-east side of Nigg Bay (Cromarty Firth) and the grazing marshes of the Beauly Firth, Munlochy Bay and the Loch of Strathbeg are of recognised importance for wintering greylag goose and pink-footed goose (*Anser brachyrhynchus*).

11.5 Human activities and conservation

As with all the habitats considered so far, traditionally managed coastal grazing marsh has been further modified or destroyed by a variety of human activities, the most important of which are listed below (Table 11.29).

Table 11.29. Key management issues for grazing marsh in the United Kingdom

HABITAT LOSS FROM:	**ASPECTS OF ENGINEERING:**
Drainage and conversion to arable land;	Sea defence structures;
Other agricultural intensification;	Remedial engineering;
Building roads and other infrastructure including	Climate and sea level change;
industrial development, tipping.	The importance of storms;
AGRICULTURAL IMPACTS:	Saline intrusion.
Eutrophication;	**MANAGEMENT:**
Change in water levels.	Overgrazing/undergrazing;
OTHER ISSUES:	Maintenance of grazing regimes;
Invasive/alien species;	Maintenance of water regimes.
Pipe-laying, including gas and oil.	

The on-going and traditional management are the keys to the survival of the nature conservation interest of coastal grazing marshes. This usually involves low-intensity grazing by cattle or sheep and/or other agricultural use with low input fertiliser applications Those semi-permanent grasslands and drainage ditches which become flooded in winter also depend on the maintenance of traditional hydrological management. These provide open water suitable for a variety of wintering waterfowl and dry tussocky grassland for breeding birds in the summer. Traditional management of the drainage ditches (involving periodic cleaning) is also a key to the maintenance of their flora and fauna. Some of the principal conservation issues influencing this habitat are highlighted below

11.5.1 Arable intensification

The sequence of primary enclosure and subsequent improvement for intensive arable land is discussed in chapter 5 for grazing marshes derived from saltmarsh. The key to the survival of grasslands and associated habitats is freedom from the second stage of conversion, that to intensive agriculture. The extent to which losses have occurred through this activity has been well documented only in the UK, particularly in southeast England (Thornton & Kite 1990). Drainage and subsequent ploughing results in all but the most common plants and animals being eliminated from the system. Often only a few ditches remain. These soon become impoverished and choked with vegetation, as fertiliser run-off changes the nature and diversity of the communities in the ditches.

The impact of the activity became significant during the 1960s, and earlier, when large scale agricultural 'improvement', including conversion of permanent grasslands for use as arable land, began. The scale of loss on the coast is shown by reference to the extent of marshes surviving in the 1930s. Since then between 30-70% have been lost in various parts of southeast England (Thornton & Kite 1990). In one area in Broadland, an extensive area of coastal wetland in eastern England, Driscoll (1983) found that 33.5% of drainage ditches had been lost due to infilling between 1971 and 1973, while many of those remaining had become floristically impoverished. A similar picture emerges further north with the major area of coastal marshes on land claimed in the Grampian Region of Scotland adjacent to the Loch of Strathbeg. Much of this land was converted to improved pasture or arable use with a few areas still managed for grazing.

11.5.2 Infrastructure

Loss to infrastructure development is less frequent, though a few individual cases can be cited. In the early 1990s marshes around the Thames estuary were variously threatened from roads, a power station and tipping of dredged spoil. On the Gwent levels in south Wales road construction has also caused some losses. Paradoxically, inundation from the sea as part of a scheme to mitigate the effects on bird life of the amenity barrage across Cardiff Bay (part of the Severn Estuary) will destroy other areas of interest.

11.6 Management

As has been stated above, coastal grazing marsh is essentially a habitat created by human use and one which requires continued (traditional) management for the maintenance of its nature conservation interest. Grazing by domestic stock is, as the name implies, the primary management tool. In the absence of high levels of artificial fertiliser application, this sustains the flora and fauna, including breeding birds. In addition control of water levels is also important for wintering waterfowl and the fauna of the drainage dykes and ditches.

11.6.1 Grazing regimes

In keeping with other coastal habitats used for grazing by domestic stock it is the density and period of grazing which help determines the conservation interest. At low levels tussocky grasslands provide nesting habitat for a variety of breeding birds and invertebrates. Inputs of artificial fertiliser accompanied by higher stock densities decreases the plant diversity and vegetation structure, resulting in lower numbers of breeding birds and invertebrates. Although slightly higher stock densities may be appropriate for the maintenance of the conservation interest of coastal wetlands than those recommended for saltmarshes (Chapter 5), the principles can also be applied to this habitat. Knock-on effects of higher stock levels include eutrophication of the drainage ditches which may reduce the plant diversity and associated invertebrate fauna.

11.6.2 Hydrology

Maintaining water levels appropriate for agricultural use and at the same time conserving biodiversity is also a key issue. Conflicts arise when the desire to increase stock use involves extending the grazing period. A farmer requires low winter, and high summer water levels to provide the best grazing. In contrast nature conservation management requires high winter and lower summer water levels. This regime caters for wintering waterfowl and in spring and summer breeding by a variety of wetland birds. The value of a regime favouring wildlife is illustrated at Holkham nature reserve on the coast of North Norfolk. By creating higher water levels throughout the year a dry grassland was transformed in to a mosaic of grass fields, reedbeds, and open water. This resulted in an increase in wintering and breeding birds which attained nationally important status (Harold 1995).

On some areas of ‘improved’ and more intensively used coastal grazing marsh the remaining drainage ditches can retain a rich flora and fauna. However, a combination of drainage of the surrounding land, nutrient enrichment and reduced ditch management can cause ditches to dry out or become clogged with vegetation (Doarks 1990). For the drainage ditches to retain their nature conservation interest, especially that associated with the invertebrate fauna, stable water levels and regular maintenance are required. This maintenance should aim to create as much habitat variety as possible by rotating both the scale and periodicity of cleaning operations.

11.7 Sea defence

It is clear that having established agricultural land on former tidal land a desire to protect it is a major preoccupation. Where earth banks enclose the marsh these have their own conservation interest, providing a refuge for a number of rare plants and animals (notably invertebrates) of transitional and upper saltmarsh. Some of the implications of their management for sea defence are discussed below.

11.7.1 Maintaining sea-walls

Many existing enclosures have been built up as earth banks made from material excavated from the saltmarsh, as is the case at many sites on the east coast of England. The methods subsequently used for maintaining them and increasing their height, have involved the excavation of more material either from the marsh immediately to seaward of the bank, or inland (Figure 11.85). In areas where extensive enclosure has already taken place and where saltmarsh survives this involves destruction of the highest and often most diverse zone on the marsh. In the case of the Wash, where systematic raising of the sea banks took place in the 1980s, it has been estimated that the total area of marsh affected in this way amounted to about 30-40ha. This area is equivalent to the average annual rate of historic enclosure at the same site (Doody 1987). In some areas in northwest England and the German Wadden Sea, turf cut from the adjacent marsh is also used to cap the bank. This causes a further loss of marsh vegetation at least in the short term.

Figure 11.85. Continuing the 'battle' with the sea; raising sea walls in Essex. Note the level of the sea outside the sea wall which is higher than the land in the foreground

11.7.2 Grazing (of sea walls) and impact on scarce plants

Some of the older earth banks, particularly in north Kent and Essex, southeast England provide refuges for several upper saltmarsh species such as *Artemisia maritima*, *Hordeum murinum* and *Bupleurum tenuissimum*. Grazing is an important factor for their survival. Although high stock numbers can be detrimental to the more sensitive species, the absence of grazing can be equally if not more destructive. Vigorous species of grass, such as cocksfoot *Dactylis glomerata* as well as scrub, may also endanger survival and may need to be controlled.

11.7.3 Other issues

There is some evidence to suggest that the agricultural value of newly created arable land derived from saltmarsh is not without its problems. In the permanent pastures of undrained coastal grazing marsh a greater proportion of sodium is present in the soil water. Drainage and cultivation of these soils has led to a transfer of sodium from the soil water to the soil particles. This has led, for example, in areas of North Kent to deflocculation of the soil which has in turn caused failure of the drainage system. A study of this situation by the Soil Survey of the UK resulted in the conclusion that "this land cannot be regarded as particularly suitable for arable cultivation" (Hazelden et al. 1986). It is ironic that a semi-natural system, which has developed over the centuries in harmony with human use, is destroyed in two decades (Thornton & Kite 1990) with what appears may have been of doubtful long-term agricultural benefit.

11.7.4 Set back ('managed retreat')

The 'saltmarsh squeeze' described in Chapter 5 may result in a desire to give up the most seaward sea wall as a means of improving the life of the coastal defences. In this way the landward limits of the coast are 'set back'. Where this results in loss of intensively farmed arable land, the result is likely to have positive benefits for nature conservation (Doody 1996). The position is different where the landward habitat is grazing marsh. Here there is a potential conflict with the maintenance of the high conservation interest of the existing habitat. On the one hand the conservationist may applaud the decision to "work with nature" under the former scenario but may be less inclined to accept losses to habitats considered precious. In areas of relative sea level rise, the desire to allow the sea to reclaim enclosed tidal land may, on economic grounds be deemed to be more acceptable on land of low agricultural value. Unfortunately these sites are also often areas of conservation interest. Under these circumstances the conservationist may have to reconsider the traditional approach to protecting important sites and accept that some change is inevitable. In this context coastal grazing marsh which has been largely created by human intervention may acquire an alternative interest. The dilemmas highlighted by this situation are considered in more detail in the final chapter.

11.8 Salinas as secondary habitats?

Salinas (salines) are seldom described in textbooks covering coastal habitats. However, in some parts of the world they represent an important wildlife resource. As with coastal grazing marshes, the combination of traditional low-level human activity with the natural assets of the system has led to the development of their own special nature conservation interest. The use of coastal lagoons for salt production is a widespread activity especially around the Mediterranean. This ranges from small artisanal sites to large scale commercial activities (Walmsley 1991). At some of the larger sites more intensive salt production takes place which can reduce the wildlife interest. Even here commercial operations with the right management can host endangered flora (including orchids) and fauna such as the flamingo (*Phoenicopterus ruber roseus*), shelduck and avocet (*Recurvirostra avosetta*), showing how economic interests can benefit wildlife (Dugan 1993, Sadoul et al. 1998). Although they are an important component of many deltas they are here described separately because of the essentially artificial origins of many, and the importance of historical management to the maintenance of their nature conservation interest.

11.8.1 Salinas, origins and definition

Natural salt lakes and saltpans occur where shallow water evaporates in areas where the climate becomes very hot. They occur in desert areas in North Africa and inland in dry continental areas such as central Turkey. Demand for salt has resulted in many artificial salinas of various sizes. The warm dry summers of the Mediterranean are particularly suitable for salt production.

The production of salt from coastal salinas has taken place in northwest Europe from at least Roman times and in the Mediterranean probably much earlier. Evidence exists from places around the Wash and Essex coasts in southern England of ancient saltpans excavated in saltmarshes (Simmons 1980). In Mediterranean salinas, seawater moves through a series of lagoons where salinity increases until saturation is reached (>300 g l^{-1}). At this stage salt crystals precipitate down through the water column and form a thick salt crust on the bottom.

11.8.2 Defining the habitat

Salinas may be defined as being any area of wetland where hypersaline conditions facilitate the precipitation of salt. These may be natural or man-made and include 6 distinctive forms as defined by Walmsley (1995):

- Natural Salt Lakes
- Primitive Saltpans
- Artisanal Saltpans
- Artisanal Salinas
- Industrial Salinas
- Non-productive/abandoned Salinas

In addition to the large natural salt lakes, salinas range from the small primitive and artisanal saltpans excavated from rocks and flat coastal sediments respectively, to larger artisanal and industrial salinas. Each has its own nature conservation interest which includes salt-tolerant plants and invertebrates, breeding and over-wintering seabirds and shorebirds, including waders, and features associated with relatively remote undeveloped land. In their natural form salinas find their most extensive expression in the salt lakes and depressions of the arid and semi-arid areas of North Africa (called "sebkhas" & "chotts") and found extensively inland in Algeria and Tunisia (Walmsley & Duncan 1993). The distribution of salt-producing salinas in the Mediterranean is shown in Figure 11.86 below (after Sadoul et al. 1998).

The annual mountains of white salt are testimony to the industrial scale of production at sites, such as the salina of Salin-de-Giraud (12,000ha), in southern France. This, the largest European salina, produces up to one million tonnes of salt per year (Britton & Johnson 1987). The Narta Lagoon in Albania (4,110ha) has, in contrast, a production of only 120,000 tons of salt per annum (Gjiknuri & Hoda 1996).

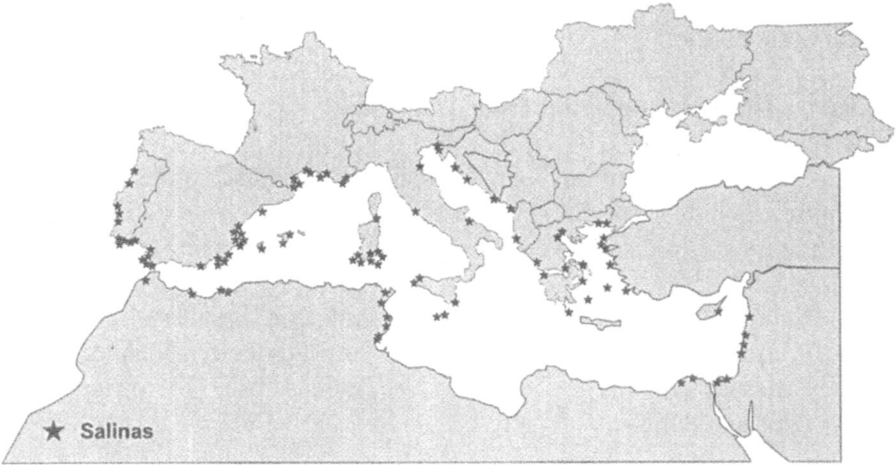

Figure 11.86. Present distribution of salinas in the Mediterranean

11.8.3 Nature conservation importance

It is perhaps surprising, given the origin of many of the major salinas from artificially enclosed lagoon systems, and the hypersaline conditions which exist, that they have a high nature conservation value. Some of these ecosystems are, in fact, wetlands of international importance with a rich and diverse flora and fauna. They host important breeding populations of aquatic birds, as well as rare and endangered species.

11.8.4 Birds

The most spectacular manifestation of the importance of salinas for birds is the presence of the flamingo. A so-called "Flamingo Triangle" stretches from the salt lakes of north Africa where the birds overwinter, to the artificial salinas of the Camargue and Ebro deltas and the more natural Marismas wetland in the National Park of Doñana in the western Mediterranean. In addition the more northerly sites are vitally important resting and refuelling sites for many thousands of waterfowl and shorebirds migrating between the Palaearctic breeding grounds and their winter quarters in Africa (Walmsley & Duncan 1993). The winter avifauna is less diverse, but for those species that do overwinter the habitat is also vitally important.

Birds present in Mediterranean salinas include gulls (Laridae), terns (Sternidae), shorebirds (Limicolae) and other aquatic species: Ardeidae (herons), wildfowl (Anatidae) and flamingos. A list of key indicator species of aquatic birds in salinas, together with their status, are shown in Table 11.30.

Table 11.30. A list of important species in Mediterranean salinas, Walmsley (Pers. comm.)

Gulls and Terns		Status
Yellow-legged Gull	*Larus cachinnans*	C *
Black-headed Gull	*Larus ridibundus*	C *
Mediterranean Gull	*Larus melanocephalus*	R/E
Slender-billed Gull	*Larus genei*	R/E
Audouin's Gull	*Larus audouinii*	R/E
Common Tern	*Sterna hirundo*	C
Gull-billed Tern	*Sterna nilotica*	R/E
Sandwich Tern	*Sterna sandvicensis*	R/E
Little Tern	*Sterna albifrons*	R/E
Shorebirds		
Oystercatcher	*Haematopus ostralegus*	C L
Kentish Plover	*Charadrius alexandrinus*	C L
Little-ringed Plover	*Charadrius dubius*	R
Redshark	*Tringa totanus*	R
Avocet	*Recurvirostra avosetta*	R/E
Black-winged Stilt	*Himantopus himantopus*	R/E
Stone Curlew	*Burhinus oedicnemus*	R/E
Water Rail	*Rallus aquaticus*	R
Other aquatic species		
Little Egret	*Egretta garzetta*	C L
Cattle Egret	*Bubulcus ibis*	C L
Grey Heron	*Ardea cinerea*	C L
Flamingo	*Phoenicopterus ruber roseus*	C
Shelduck	*Tadorna tadorna*	C L
Mallard	*Anas platyrhynchos*	C L
Red-crested Pochard	*Netta rufina*	R/E

Legend:

C = Common

C L = Common locally

C * = Common/ problem species

R = Rare

R/E = Rare and endangered species

11.8.5 Plant communities

Only a few halophytic species can survive in the saline to hypersaline conditions of salinas. Amongst these are *Salicornia perennis*, *Suaeda vera* and *Atriplex portulacoides*, species ranging northwards to the saltmarshes of southeast England. Around the margins of some of the older, smaller traditional and abandoned salinas a variety of other species may be present. *Tamarix* spp. (tamarisk) and *Juniperus* spp. are amongst the more typical shrubs and *Pinus pinea* may be found where old dunes still survive. The saline waters themselves have very few species, with *Ruppia maritima* and the alga *Enteromorpha* characteristic and often dominant.

11.8.6 Invertebrates

Invertebrates are very few in species though high in density, especially during the summer months. The species which are present vary between individual salinas with water salinity being of over-riding importance. Population changes are very rapid with one species absent on one day but present in abundance a few days later. The majority of species occur in the 40 g l^{-1} - 150 g l^{-1} salinity range, remaining relatively stable between 70-150 g l^{-1}. Above this species of prey begin to die out and with them their predators (Sadoul et al. 1998). Above 250 g l^{-1} only *Artemia* spp. (brine shrimp) is present; it forms an essential food resource for many species of birds using the habitat.

11.9 Management

Although salinas, as defined here, have been largely created by human intervention their management is not always benign. 'Natural' deltas behave in an erratic way, erosion and accretion creating and destroying bird breeding sites as environmental conditions change. The conversion to salinas creates stable water levels, which result in erosion of breeding sites. As the salinas complexes expanded, for example in the Camargue, species such as slender-billed gulls, avocets and terns continued to use the area as new breeding sites were created. When the salinas stopped expanding existing smaller islets continued to be lost through erosion, without new ones being created. Larger islands became colonised by predatory herring gulls and the other less robust species could not compete effectively and their numbers declined (Sadoul et al. 1998). Opportunities exist for remedial action to be taken on industrial sites, but abandoning this use can be as devastating as conversion to other activities such as fish-farming (see below).

11.9.1 Industrial salinas

To some extent the scale and intensity of use determine the conservation interest of an individual saline. The more intensive, efficient, and presumably money generating, the more the natural interests are compromised. However in some respects salinas differ from most other coastal habitats. Because of the very specialised nature of the biological interest and the long history of human association, opportunities are provided for practical management. Amongst these the presence of suitable breeding habitat (small islets) for waterbirds, notably flamingo and shelduck, are of particular note.

Artificial islands for nesting birds and decoys (to attract terns and gulls) are the most practical forms of management (Walmsley 1995) assuming that the water management is conducive to the presence of birds in the first place. There are only a few recognised nature reserves in industrial salinas of which the Ebro Delta (Figure 11.87), the Marismas wetland, southern Spain and the Camargue in the Rhone Delta are well known. In the last of these the largest breeding population of flamingo in the Mediterranean survives in specially protected areas within the commercial salt-production area. Since 1974 the birds have bred on specially created islands within otherwise undisturbed lagoons which also have artificially circulated sea water (Johnson 1982). The high water levels required for salt production in the summer coincide with the main breeding period between March and August (Johnson 1992).

Figure 11.87. Flamingos are protected in some industrial salinas; the Ebro Delta, Spain

11.9.2 Abandoned salinas

Many of the smaller salinas where salt was harvested manually have been amalgamated into larger ones where machines do the work. Abandonment is the result of modernisation and economics, and where smaller salinas cannot be amalgamated or improved to produce the maximum tonnage of salt per annum. These sites can be of considerable potential value for conservation as water level management is possible which can improve feeding areas and breeding sites for birds. Actively controlling salinities (to keep levels < 240 $g\ l^{-1}$) improves productivity. At the same time manipulating water levels such that there are a range of conditions (dry for several years, recently flooded, stable and combinations of these) can be effective in promoting breeding of species such as ground-nesting plovers (Sadoul et al. 1998).

11.10 Fish-farms and rice fields

A cause of decline in nature conservation importance is that associated with the change in use of salinas to fish-farms. Evidence from Portugal shows that in the case of the latter, using black-winged stilt (*Himantopus himantopus*) as an indicator species, only 3% of breeding birds use fish-farms whereas 69% use salinas (Rufino & Neves 1992). In addition, in areas with fish-farms herons and other fish-eating birds are seen as a threat to commercial activity and birds are either frightened away or shot (Walmsley 1994).

11.10.1 Rice fields

Many lagoons in the Mediterranean are cultivated for rice (Figure 11.88). The requirement for water table manipulation has led to a sequence of habitats which also has value for wetland birds. The early flooding (by comparison with natural wetlands where autumn rainfall is more usual) helps to make these artificial areas produce suitable prey earlier in the year for a variety of wetland birds. In the Ebro Delta in Spain this interest is very high. A combination of traditional rice cultivation adjacent to more natural lagoons and other coastal habitats has resulted in a wide range of important migratory and breeding wetland birds being present. Some Portuguese salinas have also been converted to rice fields. But as salt production has declined so have the bird populations (Rufino & Neves 1992). Again the result of this sequence of progressively more intensive use, ultimately excludes the wildlife interest almost completely and any vestige of natural habitat is destroyed.

Figure 11.88. Mechanised rice cultivation, Ebro delta, Spain

11.10.2 Hydrology

The key factor in the maintenance of these artificial wetlands is the timing and level of flooding. Birds, particularly herons, using rice fields and other freshwater coastal wetlands, depend on the level of water being maintained throughout the summer. Changes in irrigation regimes, use of pesticides, and fertiliser applications can all impact on their value. Careful consideration of these issues is a key to retaining these important areas and their wildlife (Hafner & Fasola 1992). The requirement here, as in so many of these wetland areas, is a better understanding of the water management and its effect on bird populations. Since all these areas are essentially artificial habitats, this can be achieved only when there is close co-operation between the wetland managers and those interested in wildlife conservation.

11.10.3 Conservation

The future of both coastal grazing marshes and salinas depends, at present, mostly on the continued economic justification for each activity. As economic circumstances change and the drive to more efficiency occurs, traditional management of these habitats tends to stop. In all but a few cases, where nature reserves or other subsidised management activity takes place, wildlife interest is also lost. Since abandonment, at least in the case of salinas, can be as damaging to wildlife interests as more intensive use or conversion to aquaculture, active management either to continue salt production or to maintain the interests of wildlife is required.

This can be achieved in only a few cases when nature reserves are established (see for example, Parc Natural S'Albufera on the Balearic island of Majorca, Figure 11.89). Here a coastal wetland, including a relatively small area of saltpans (operated from 1946 and abandoned in 1977), is actively managed as part of a nature reserve. For many of the other sites subsidies to retain traditional management practices are needed. Without this financial commitment, the voluntary conservation organisations most likely to undertake the required level of management will be unable to do so owing to a lack of resources.

Figure 11.89. S'Albufera, Parc Natural, Majorca. A complex coastal wetland including abandoned salinas

12. ECOLOGICAL NETWORKS

12.1 Introduction and scope

Mobile species may rely on one or more habitats during the course of their life cycle and these may be spatially separated by a few metres or several thousand kilometres. Previous chapters have described the important spatial and temporal relationships between and within coastal habitats. These include sometimes complex ecological and geomorphological links where external forces may force rapid change. The management of an individual site can also fundamentally alter the nature of the component habitats and all these factors can influence the value of the area to visiting species. When considering the conservation of a mobile species it is important, therefore, to understand the individual habitat and feeding requirements at the different stages in its life cycle. This is particularly important for those species with special needs when breeding (e.g. sea cliffs for sea birds, shallow dune slacks for natterjack toad or saltmarsh with a good structural diversity for redshank), or feeding habitat (e.g. estuary mudflats for wintering waders or saline lagoons for the flamingo). Those species which are more catholic in their tastes are less dependent on specific habitats and therefore more resistant to change.

Loss of part or all of a habitat to development and the implications of this for the conservation of migratory birds using the site as part of the network upon which the species depends is a common conservation issue. Piecemeal agricultural encroachment, also affecting drainage patterns and the ability of some areas to accommodate storms and high water levels, is also a feature of many coastal wetlands. Damming of rivers reduces sediment load and can result in erosion at the margins causing further loss of habitat. These changes, which have been chronicled in the preceding chapters, have often resulted in the many classic conflicts between 'developers' and 'conservationists'. This is particularly true when proposed major developments affect key elements in bird migration networks.

There is a relatively good understanding of the migration patterns of many waterbird species. Studies, co-ordinated by organisations such as Wetlands International, include those showing migration patterns (Scott & Rose 1996) and a comprehensive account of the status, trends, distribution, ecology and conservation management needs of geese in the Palaearctic (Anon 1999). These and other publications provide essential information, which can facilitate agreement amongst different countries along migratory pathways. Despite this wealth of information, the habitat preferences and hence conservation requirements at the different stages in the life cycles of many groups of bird species are still often lacking. The situation is even more difficult for the marine environment where only very broad patterns of distribution have been elaborated and for a very few species (some whales and sea turtles). Thus although it is possible to articulate the concept of networks in terms of patterns of migration, it is much more difficult to provide a coherent and detailed picture of the management needs of individual mobile species which travel between a range of different habitats. This chapter attempts to provide an insight into these issues and to analyse the implications of habitat loss or change resulting from management decisions. It also looks at the way in which habitat restoration can play a part in the conservation of migratory species. The issues are considered at international, regional and national scales, as well as, 'ecological networks' within sites or habitats. It is based on a preliminary assessment of the ecological components of a coastal and marine network prepared within the framework of the European Ecological Network (Bennett 1991) which is part of a pan-European Biological and Landscape Diversity Strategy (Anon 1996b).

12.1.1 Protecting migratory birds

The concept of developing a network of sites to protect nature conservation interests is not new. Both national and international designations have sought to provide for the identification and protection of a series of sites to ensure the conservation of certain migratory species across their range. In the Americas there are two main pathways, the Pacific and Atlantic flyways. A number of conservation sites along these corridors provide feeding and resting areas for the autumn and spring migration of ducks, geese and other birds. In the USA conserving migrating birds has long been a part of the rationale for the establishment of nature reserves and is of particular relevance to the National Wildlife Refuge system, where many sites are managed especially for waterfowl such as the canada goose (Figure 12.90).

Today in the USA up to one billion US$ per annum are estimated to be spent on the conservation of migratory birds. Initiatives to unite the flyways of the Americas include the North American Waterfowl Management Plan and the proposed linkage of North and South America being promoted by Wetlands International amongst others (Davidson 1999). Similar initiatives exist in Europe for the East Atlantic Flyway and recently an African-Eurasian Migratory Waterbird Agreement came into force under the Bonn Convention on the Conservation of Migratory Species of Wild Animals. This is legally binding on the signatories to the convention (49 world-wide by May 1996) and provides for the conservation and effective management of migratory species across their range including protection for a variety of rare and threatened species such as sea turtles and the monk seal (*Monachus monachus*). Agreements under the Convention also cover cetaceans (Agreement on the Conservation of Small Cetaceans in the Baltic and North Seas 1991, ASCOBANS and Agreement on the Conservation of Small Cetaceans of the Black Sea and Mediterranean Sea and Contiguous Atlantic Area, ACCOBAMS). Together with other agreements and conventions there is a developing world approach to the conservation of a number of groups of specialist migratory animals.

Figure 12.90. The Canada goose is dependant, during its migration, on a sequence of sites such as those in the National Wildlife Refuge system in the USA

12.1.2 Pan-European Biological and Landscape Diversity Strategy

The 'Pan-European Biological and Landscape Diversity Strategy' is a European response to support the implementation of the Convention on Biological Diversity, agreed at the Earth Summit in Rio in 1992 (UNEP 1992). It "..introduces a co-ordinating and unifying framework for strengthening and building on existing initiatives" across Europe. Over the next 20 years its aim is to include biological and landscape diversity considerations into social and economic sectors by integrating them with the main human activities (Nowicki et al. 1996, Nowicki 1998, Council of Europe 1998). Within this strategy the establishment of a European Coastal and Marine Ecological Network (ECMEN) as part of the European Ecological Network (1996-2000) has been elaborated (Doody & Salman 1998). This chapter now deals with the key elements required when adopting a network approach to coastal conservation based on some of the examples used in the ECMEN review.

12.2 The nature of coastal networks

Networks are defined by the species that depend on them for their survival. The America flyways stretch for some species from Alaska to the southern tip of South America and give an indication of just how wide-ranging their distribution can be. Individual species such as Arctic tern show enormous migration ranges, in this case stretching from its breeding grounds in the far north to the southern Antarctic Ocean (Meade 1983). The migration paths of these and other species, including some wading birds, are more extensive than for whales, see for example the humpback whale (Evans 1990 and Figure 12.91).

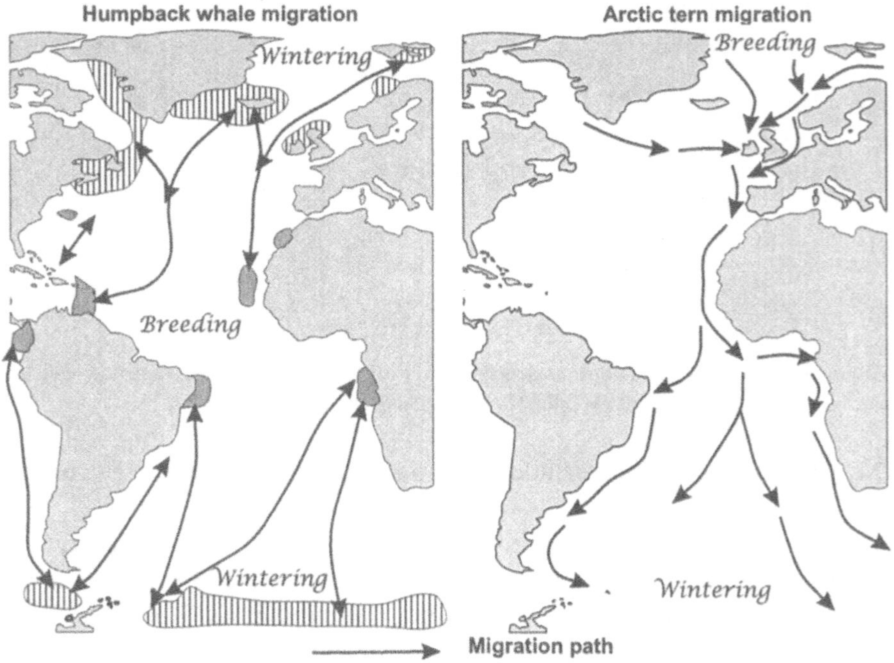

Figure 12.91. Migration paths for humpback whale and Arctic tern after Evans (1990) and Meade (1983)

The survival of many species depends on a combination of '**core areas**', usually used for breeding purposes or feeding areas where the species has specialist requirements. The concept of '**stepping stones**' is appropriate for some birds such as the knot and barnacle goose, though less so for marine mammals (seals and cetaceans) where more widespread dispersal takes place outside the breeding season. For some fish (e.g. salmon) the notion that there are '**corridors**' within which species may be particularly vulnerable is also important. Rivers, and in the marine environment straits between islands or fronts between warm and cold water, can also have a particular significance. Air corridors are less obvious and more difficult to define but many migrating birds depend on them. Also important are the narrow straits between continents and seas (Gibraltar and the Bosphorus are two well known examples) where birds (notably birds of prey) gather to make best use of thermal currents to carry them across open water.

12.3 Networks, some examples

The migration patterns of a number of species have been used to develop and illustrate the elements of a European Coastal & Marine Ecological Network (Doody & Salman 1998). These are in no way comprehensive but include waterbirds, marine mammals and fish. The descriptions which follow attempt to show something of the main types of networks which exist in coastal locations and the different habitat requirements for individual species.

12.3.1 'Stepping stones' - barnacle goose (*Branta leucopsis*)

The barnacle goose breeds in the high Arctic in restricted areas where its habitat preference is related, in part at least, to freedom from predation by arctic foxes. The species migrates along set paths to its wintering areas stopping en-route at traditional locations ('**stepping stones**'). Wintering sites are also highly localised (Figure 12.91).

In recent years reduction in grazing of livestock on the smaller islands, which are used to refuel during migration, has diminished the suitability of pastures and increasingly individuals stray onto agricultural grasslands nearby (Scott & Rose 1996). Feeding preferences in the winter include *Zostera* spp. (marine plants of the intertidal and sub-tidal estuarine mudflats) and other plants of saltmarshes, though once this food source is depleted individuals will move to coastal agricultural pastures giving rise to conflict with farmers.

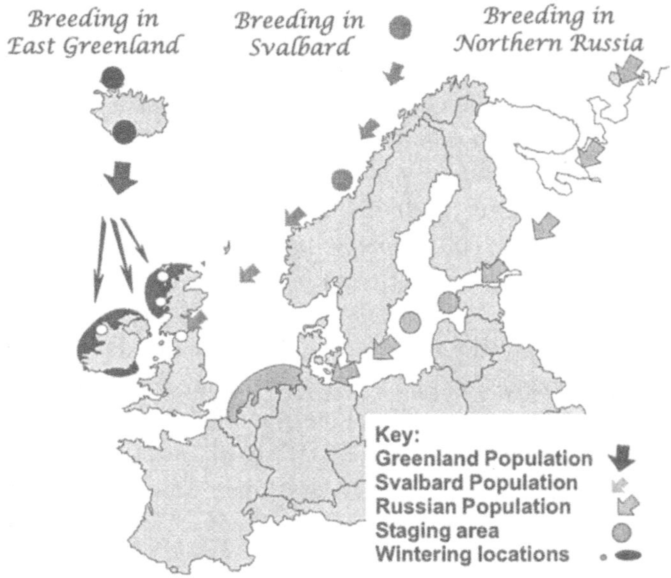

Figure 12.92. Autumn migration routes and 'stepping stones' for wintering barnacle goose

12.3.2 'Core' breeding areas, dispersed migration and feeding areas - marine mammals and sea turtles

Seals, sea turtles and cliff-nesting seabirds depend on specific and sometimes very limited '**core**' areas on land for breeding. Seals are representative of marine mammals which are widely dispersed and spend most of their lives at sea, in this case feeding on pelagic species. They require sandy banks, other sedimentary or rocky shores and islands to haul out for resting and breeding. Loggerhead turtles (*Carretta carretta*) are good swimmers spending almost their entire life in the sea. When nesting, their eggs need air and they return to the same stretch of coastline every breeding year and from June to September each female lays between 80-100 eggs in sand above the high water mark. This species and other sea turtles are especially susceptible to damage to their nesting sites. They highlight the links between marine and coastal environments because of their dependence on the sea for food and the need for quiet undisturbed places to breed.

12.3.3 'Core' breeding areas, dispersed feeding and migration - Eleonora's falcon

Eleonora's falcon has a pattern of migration, which takes the species between two very distinct geographical zones. It winters in Madagascar where it is widely dispersed, feeding on a range of insects and in a wide variety of habitats. These include coastal cliffs and cultivated areas, such as rice fields, lightly wooded slopes and coastal plains (Walter 1979). It is a small falcon, which feeds mostly on small birds and hence does not require specialist habitats or 'stepping stones'. Unlike soaring birds of prey its powerful flight does not require updrafts to be able to sustain travel over the considerable distances between Madagascar and its breeding areas on the Mediterranean and nearby Atlantic coast, where most of the world population is located (Meade 1983). It is a colonial breeder inhabiting steep rocky coastal cliffs and islands, especially along the coast of Greece, though extending as far west as the Canary Islands. Its breeding season is later than other species and timed to coincide with passerine migrations ensuring a supply of food for its young. During this period of late summer it congregates in specific geographical areas where prey are particularly plentiful (Walter 1979).

12.3.4 Dispersed breeding and winter feeding

Many species, especially smaller migrants such as warblers, wagtails and larks, do not rely on specific and readily identifiable geographical locations or specialist habitats. For many species a combination of habitats located at short intervals along the migration path is more important than specialised 'stepping stones'. Their smaller body weight results in migration paths being followed, possibly in a series of 'hops'. Shore lark (*Eremophila alpestris flava*) is a rare but regular winter visitor to the coast of Britain, where it feeds on seeds almost exclusively at the margins of saltmarshes and dune strandlines in East Anglia. In Germany and Russia it also feeds on grain and other seeds and in a variety of habitats (Simms 1992). Its migration route lies partly over land, taking the shortest route across the various seas or hugging the coast as in the southern movement of the Baltic population (Figure 12.93).

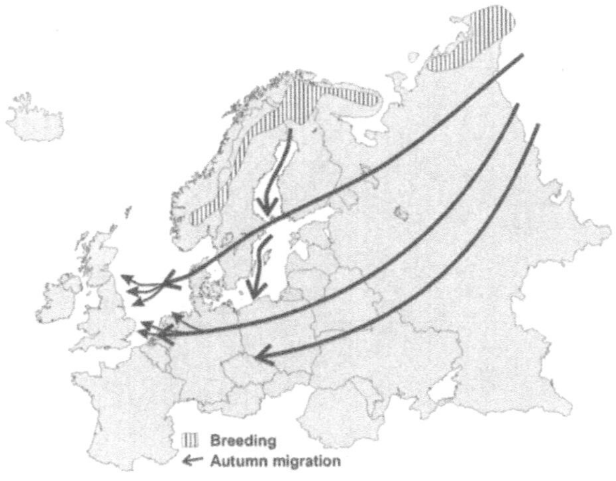

Figure 12.93. The shore lark breeds in the arctic and sub-arctic (race *flava*) and winters along coastal shores of the Baltic and southern North Sea (after Simms 1992)

12.4 Geographical scales

Bird migration studies have ensured that migration patterns of many species are, comparatively speaking, quite well known. At each stage in the life cycle there may be very specific location and habitat requirements. During migration individual species may have traditional stopping-off places which meet their needs for rest and refuelling. These wide-ranging patterns of distribution are only part of the possible network structure. To illustrate these different elements the life cycle of the red knot (*Calidris canutus islandica*) is used. There are a number of important stages in the life cycle of this species (breeding, the main migration, moulting and winter-feeding), each relying on different habitat and feeding requirements. The species has a particularly extensive migratory pattern with individuals of the subspecies *rufus* moving between its breeding grounds in Arctic North America to as far south as the southern tip of South America. The subspecies *canutus* covers a similar distance but from the Arctic to Europe and as far south as South Africa (Piersma & Davidson 1992).

12.4.1 An 'autumn network'

Figure 12.94. The 'autumn network' of the red knot, based on Davidson & Stroud (1996).

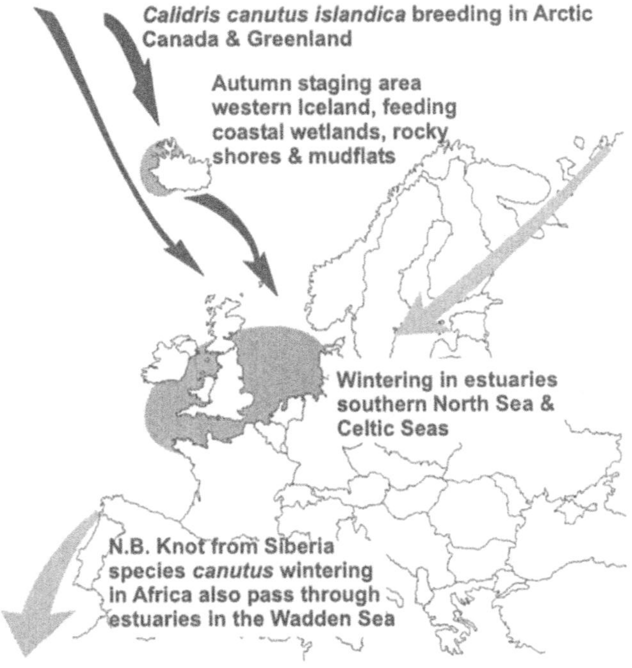

Large areas of undisturbed tundra are required for establishing breeding territory for the red knot. Upon returning to the Arctic breeding grounds nesting occurs in the summer at low densities and lasts for fewer than 2 months. After this, birds begin their return south, taking major 'steps' on their autumn migration to the southern North Sea and beyond (Figure 12.94). Their return journey also involves a series of major steps though some locations are different from those used in the spring.

12.4.2 A 'winter network'

When birds have reached the wintering areas in August, moulting (when the feathers are renewed) takes place. Birds in the Wadden Sea, where some 50% of the population occurs, tend to move westwards as the winter temperatures drop and food becomes more difficult to find. This westward movement (Figure 12.95) is part of a more general pattern. These, and other species of waders, require access to coastal shores, including tidal mudflats, with an abundance of accessible and suitable invertebrate prey. The movements continue and can occur into mid-winter or even later when climatic conditions are particularly harsh and tidal flats become frozen (Smit & Wolff 1981). Other, even more complex patterns of movement have been identified, as for example in dunlin (Pienkowski & Pienkowski 1983). This small wader moves between sites in northeast England and northern Spain or from the Wadden Sea to Portugal, western and northern France and western Britain.

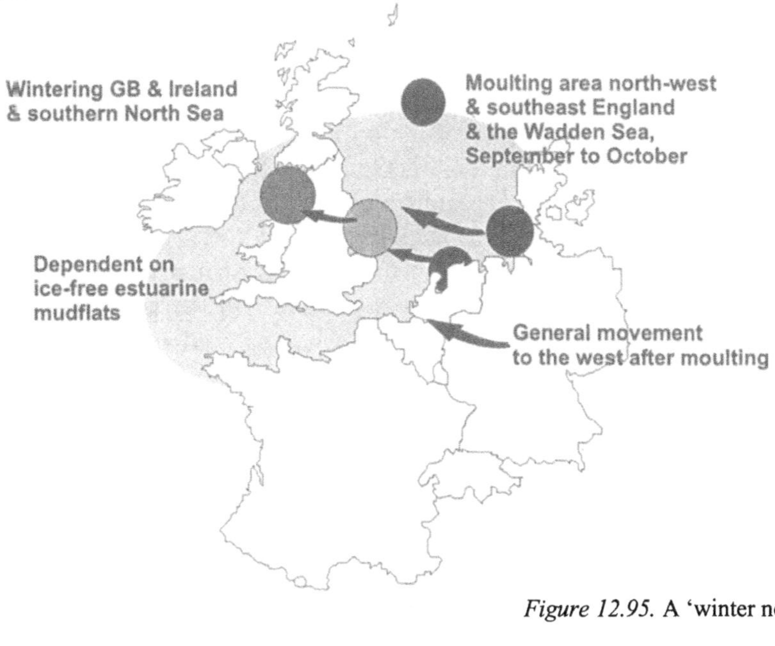

Figure 12.95. A 'winter network'

12.4.3 A 'tidal network'

In addition to the movement between breeding, moulting and winter feeding grounds, and between sites within the wintering area, individual birds also use several coastal habitats during a tidal cycle. Estuarine waders, such as the knot, typically feed on exposed mudflats, moving landward or seaward as the tide ebbs and flows. At high tide they may move to, and roost on, high level saltmarshes, sea banks or other land adjacent to the estuary. Such waterfowl may feed on farmland during the day, returning to the relative safety of saltmarsh or other estuary habitat at night. These movements are dependent on complex food-webs, nutrient pathways and energy flows upon which the ecosystem is built (Packham & Willis 1997 pp. 16-20).

12.5 The red knot, land claim and sea level rise

The above discussion has attempted to show the pattern of networks for a small number of migratory animals. In order to see how an ecological network approach could help our understanding of the conservation management of an individual species, the knot is considered in more detail.

A large proportion of the wintering knot population (including *Calidris canutus canutus* which passes through, en-route to Africa) is found in the general region of the southern North Sea including estuaries of southeast England and the Wadden Sea. Historically, these are places where extensive intertidal land has been lost through enclosure, principally for intensive agricultural use, ports and industry (Doody 1995 and Chapter 10). Land-levels are also sinking, as the southern North Sea basin continues to respond to the melting of the ice sheets which covered much of northern Europe up to some 10,000 years ago. Taken together with the general rise in global sea levels, southeast England is experiencing some of the fastest rates of relative sea level rise in Europe, which in places is in excess of 5mm per annum (Carter 1988). These forces result at the margins of many tidal areas in the tidal land being 'squeezed' by the loss of land and the rising sea level. A steepening foreshore and loss of intertidal sand and mud are a consequence.

The knot feeds on tidal mud and sand flats and the loss of intertidal areas through land-claim is especially significant. It is also vulnerable when cold weather conditions cause the habitat to freeze over. This serves to exacerbate the permanent loss of habitat with a temporary loss of feeding grounds leading to the potential for increased winter mortality. Whilst some birds may move to sites further west there may be a limit to the carrying capacity of these areas, also such that the whole population is put under greater stress.

12.5.1 Site protection and population status

The extent to which international designations have been established for the protection of major wintering sites for the knot (and other waterfowl) in Great Britain are shown in Table 12.31. All of the most important sites are protected by a variety of international and national legislative measures. These, together with the establishment of nature reserves by Non-Governmental Organisations, landscape designations, planning and management strategies, all help to protect this conservation resource. Most sites not designated or proposed for designation under international agreements (Ramsar[1], SPA[2] or SAC[3]), are Sites of Special Scientific Interest, the principal site protection legislation in Great Britain.

Table 12.31. Protected sites and numbers of the knot in Great Britain

Site Name	**Nos.**	**Ramsar**	**SPA**	**SAC**	**Threats over last 20 years**
The Wash	75000	✓	✓	✓	Land claim for agriculture
Humber Estuary	28900	✓	✓		Land claim for industry
Morecambe Bay	26300			✓	*Spartina anglica* expansion, road building
Dee Estuary	22200	✓	✓		Land claim for industry & ports
Foulness	22151		✓	✓	Airport development
Benfleet & Southend Marshes	8400	✓	✓		Refuse disposal
Inner Firth of Forth	8094				Refuse disposal
Dengie	7763	✓	✓	✓	Refuse disposal & erosion of saltmarsh
Upper Solway Flats & Marshes	5650	✓	✓		Change in saltmarsh grazing management
North Norfolk Coast & Gibraltar Point	5500	✓	✓	✓	Coast protection
Burry Inlet	5490	✓	✓	✓	Industrial cockle dredging
Thames Estuary & Marshes	4990	✓	✓		Land claim for industry
Montrose Basin	3700		✓		
Medway Estuary & Marshes	3690	✓	✓		Land claim for industry
Duddon Estuary	3603				
Teesmouth	3574				Port development
The Swale	2650	✓	✓		Land claim

1 The Ramsar Convention on Wetlands of International Importance especially as waterfowl habitat. Contracting parties are national governments, which are required to designate wetlands of international importance on a global scale and to promote their conservation and 'wise use' Ramsar sites are thus designated for their waterfowl populations, important plant and animal assemblages and wetland interest or a combination of these;

2 The EC Directive on the Conservation of Wild Birds (The Birds Directive) provides for the designation of Special Protection Areas (SPAs) for the conservation of rare, vulnerable or regularly occurring migratory species of birds. Council Directive 79/409/EEC;

3 The European Union Habitats and Species Directive provides for the establishment of Special Areas of Conservation (SACs) based on a range of vegetation types, priority habitats and rare species concentrations at a European level. Council Directive 92/43/EEC.

The Table also shows some of the pressures from enclosure and other developments that have affected the sites in the past, based on the author's experience. Recent decisions to allow development in the Thames estuary, extension of port facilities at Felixstowe in Essex and on the Dee Estuary (Wales) have resulted in further land claim with losses of both the quality and quantity of habitat with the potential for a reduction in bird numbers. From this it can be deduced that statutory designation of these sites has not ensured their full protection with implications for the conservation of the knot or other birds using British estuaries (Davidson et al. 1995).

12.5.2 An accumulation of losses

A review of flyways and water bird reserve networks (Boyd & Pirot 1989) suggest that, along with a number of other species of waterfowl, there was a considerable decrease in numbers of knot (from about 609,000 in 1976 to 345,000 in 1989). Cold Arctic summers have been linked to adult deaths and poor breeding success during this period (Boyd 1992). More recently a recovery in numbers has been observed and a new estimate of 450,000 birds for the mid 1990s has been established (Kirby et al. 1999), though still well below its population in the 1970s.

Studies in the UK and Netherlands would tend to support the view that habitat loss may reduce the overall population. Evidence from the Tees Estuary in northeast England (Evans et al. 1979), and the impact of the Delta Project in the Netherlands (e.g. Schekkerman et al. 1994), both show a reduction in bird numbers as a result of land-claim. This will, at the very least, displace birds to other locations as food resources become more quickly depleted. More significantly, although the knot population appears to have recovered after its losses in the 1970s, it has stabilised at only 70% of its numbers then. The reasons for this are not clear but land-claim and deterioration in the quality of habitat at its wintering sites could be implicated in the absence of major change in breeding success.

12.5.3 Policy responses and sea level rise

The main policy response to rising sea levels, erosion and flooding has been to protect life and property. This has usually involved the maintenance of the current line of the coast by a variety of engineering methods. These have included building new defences or the reinforcement of existing ones to prevent erosion. Additionally the protection of areas from flooding, especially when a tidal surge occurs, has involved building storm surge barriers such as the one which protects the City of London (Figure 12.96). So long as the main policy response to the effects of global warming and sea level rise is to maintain the current line of defence, loss of intertidal land will continue. The conservation of the knot and other estuarine waders may depend, in part at least, on whether hard engineering structures continue to form the only method of coast protection.

In southeast England the reduction in the area of intertidal land can be reversed by retreating from the current line of sea defence (Radley 1997). A number of experimental sites have been established including one at Tollesbury on the Blackwater Estuary where a tactical retreat was engineered under the direction of the Ministry of Agriculture Fisheries and Food and the statutory nature conservation agency for England (English Nature). The aim is to recreate tidal land and at the same time provide natural sea defences (mudflats and saltmarsh) to help absorb the power of the waves which are undermining the artificial sea defences along much of this part of the coast (Figure 5.40 in Chapter 5).

Figure 12.96. The Thames barrier is raised when storm surges threaten London

The potential for improving the opportunities for waterfowl populations is considerable. The series of estuaries lying along the Essex coast, already a proposed SAC and a site of international importance, currently hold >30,000 waterfowl in winter, including 10 wader species, more than 1% of the East Atlantic Flyway (EAF). Recreating tidal land would help to stave off further losses and has the potential for considerable increase in the carrying capacity for waders. The potential areas over which such a policy could be enacted is shown in relation to land on the Essex coast where there are extensive areas of enclosed saltmarsh currently used for agriculture (Figure 12.97).

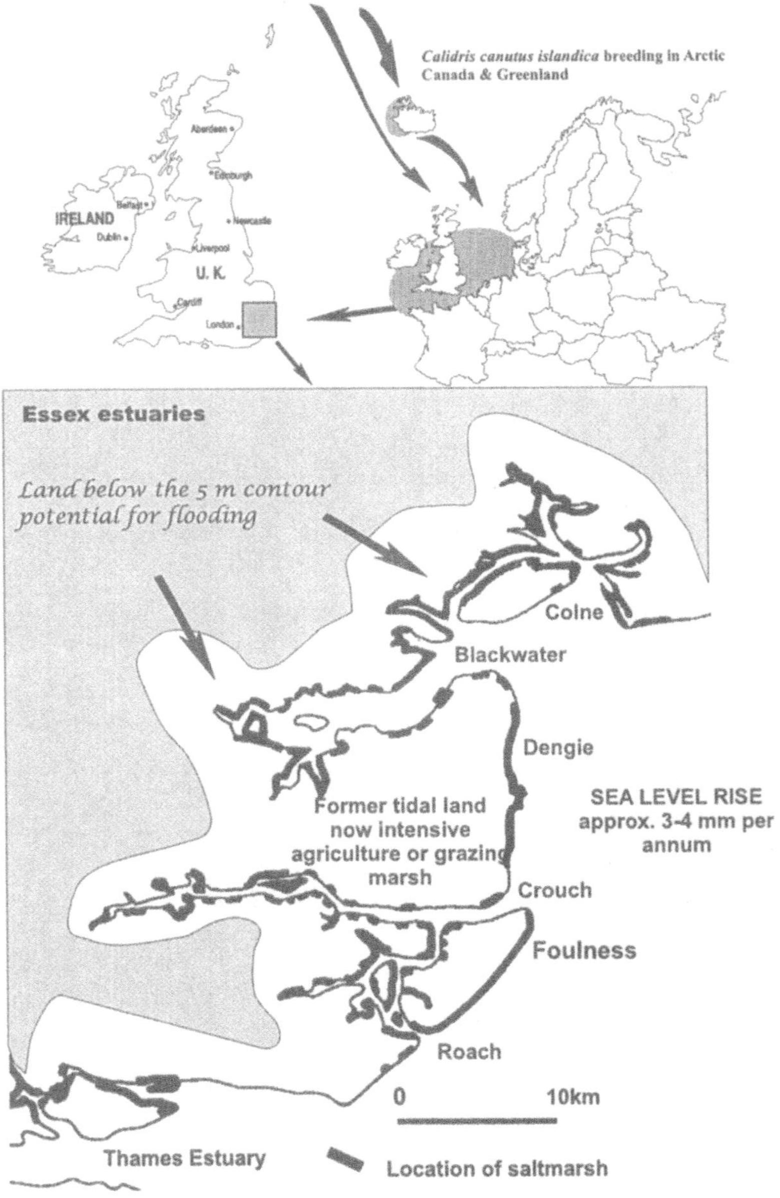

Figure 12.97. Potential areas for the recreation of tidal land on the Essex coast, England

12.6 International protection and local management

The case of the knot and the other examples described above amply illustrate the wide-ranging and complex nature of networks of migrating species. Conserving these species requires an equally complex strategy which recognises the need for adequate protection at all stages in the life cycle. This can be achieved only through an approach which crosses political and geographical boundaries. This next section attempts to build on the discussion and unravel the ways in which wider conservation objectives can be achieved.

12.6.1 Protecting 'stepping stones'

As we have seen above, and as is true for many migratory species, some protection is afforded to most species somewhere within their geographical range. However, even where international agreements are operative, the management and protection of the sites within the country boundary are often determined by national laws. Seeking wider protection such that actions in one country complement those in another is not easy.

Taking the case of the red knot we can see that protection, or even habitat restoration, in southeast England may be undermined by actions occurring elsewhere. Returning to its breeding areas in the spring may involve equally long flights as those in the autumn. On the return journey the birds may use stop-over points in western Iceland and northern Norway (Davidson & Stroud 1996). In northern Norway they appear to be faithful to two relatively restricted areas, Balsfjord near Tromsø and Porsanerfjord some 250km to the northeast, where they remain for only a few days from the 12th-16th May when they arrive, to the 25th-28th May when they leave (Strann 1992).

Loss of these staging areas could be disastrous for the whole population. Building along these shores is a potential threat which has not been quantified, but new houses continue to be erected in and around Tromsø (personal observation 1998). The relatively short period of occupation by the migrating individuals may make statutory protection difficult, even though the areas which are used are well prescribed. More general prohibition of the continuing development along the shore, which is being enacted in a number of countries around the Baltic (Nordberg 1999), may be the only practical policy response.

12.6.2 Breeding success

Breeding success is, as already stated (Boyd 1992), subject to the vagaries of the Arctic summer. Although this is a natural influence on population size, it is uncertain what the implications of climate change might be. When taken together with habitat loss of both staging areas and wintering sites, which may also become inaccessible due to low temperatures, severe depletion of the population could follow. Little is known about the interaction of all these factors on the populations of knot let alone less well studied species.

It is also clear that the habitats making up the core areas within the network of sites used by the species, and the links (corridors) between them, are subject to a wide variety of management actions, already considered in detail in earlier chapters. A key point here is that human use can fundamentally alter the nature of a habitat or ecosystem. This not only relates to habitat loss, but also to changes in management whether it involves restoration of tidal land or a change in grazing management. The former may enhance opportunities for new suitable intertidal habitat to be created and hence favour the knot and other waders in winter. The latter, for example when applied to grazing on saltmarshes (Chapter 5), may favour wintering herbivores such as the barnacle goose, at the expense of breeding species, including some waders. In this context it is vitally important to understand the requirements of all relevant species so that appropriate management options can be identified and assessed.

12.6.3 Lessons

Despite the extensive international protective measures enacted through the various conventions, the threat to migratory populations world-wide remains. Most of the conventions apply to bird species though recently other groups of animals are covered (as under the ASCOBANS and ACCOBAMS agreements, included within the Bonn Convention, see above). Often traditional approaches to site conservation and management provide the means for the implementation of international agreements. These are not always as effective as they might be for the following reasons:

- Some species are widely dispersed or present so irregularly during parts of their life cycle that traditional approaches which rely on the protection of core areas and stepping stones are inadequate;
- Some habitats (nearshore sea areas) lie outside the scope of effective legislation;
- In many areas where appropriate legislation exists it is not fully applied;
- Many potentially damaging activities are themselves widely dispersed requiring legislation and compliance, which crosses nation states and is difficult to apply to individual sites.

The above examples illustrate how varied the nature of a migration network can be. These range from the species which have very specific breeding and/or feeding requirements to those with a wider range of preferences. The former require sometimes very specialised habitats, the latter can survive in a much wider range of situations. This has important implications for the way in which conservation measures can be applied.

Even though the knot is a much studied species with a relatively simple migration system "the links between sites are not all understood." (Davidson & Stroud 1996). This example shows how the elements of a network can be identified by reference to the distribution and migration pattern of an individual species. Thus core areas, for breeding, resting and feeding on migration (staging areas), are as important as safe sites for moulting and feeding. Each stage of the life cycle is dependent on a particular habitat and if any of these are lost then the links in the chain can be broken. This may not be crucial to a common, widely dispersed species which will simply move to other 'suitable' areas. On the other hand it could be critical to the survival of rare and isolated populations, or species with highly specialised habitat requirements, which is often the case in the coastal and marine zone.

The illustrated case of set-back in southeast England also shows that when issues such as coastal defence or knot conservation are looked at from a wider perspective, interesting alliances can be formed. Together the requirements of the conservationist and the engineer can become a powerful alliance. This may help to fulfil important socio-economic requirements associated with coastal protection as well as contributing to the conservation of tidal land and of wintering waders (Doody 1992, 1996). These issues are considered further in Chapter 14.

13. INTEGRATED COASTAL MANAGEMENT

13.1 Introduction

Traditionally the cornerstone of the nature conservation is the protection of the best examples of particular habitats and species concentrations. Many precious wildlife areas have been destroyed or reduced considerably in size, others survive only as small 'islands' in a landscape used for intensive agriculture or otherwise highly modified. In the face of the continuing attrition of wildlife and their habitats, it is perhaps not surprising that the conservation movement continues to promote the protection of those places with special habitats and species.

This is achieved by a variety of means including the establishment of nature reserves, National Parks and other forms of statutory protection and by the voluntary conservation movement. The National Trust (NT) in the United Kingdom, through its Enterprise Neptune campaign, now owns and manages 885km of coastline in England, Wales and Northern Ireland and this is increasing year by year. The Audubon Society in the United States of America, through members of the Audubon Alliance, own reserves and sanctuaries, many on the coast. Ousin Island in the Seychelles is owned by the International Council for Bird Preservation and is a truly international nature reserve.

Taken together, the global network of protected areas (not just coastal sites) amounts to 30,000 sites, represented in 13 million km^2 or 9% of the total land area. In Europe it is estimated that 12% of the total land surface is protected to some degree (Green 1997). However, despite the number and scale of these protected conservation areas, damage to the environment continues.

On a global scale losses have been enormous. Although it is impossible to give an overall assessment, in the USA for example, it is estimated that two million acres of productive coastal water and marsh was lost (more than $^1/_4$ of the total) in only 32 years (Teal & Teal 1969). Hinrichsen (1998) gives a total figure of 50% loss of coastal wetlands. In the Mediterranean up to one million ha of coastal wetlands have been destroyed (UNEP Blue Plan, Grenon & Batisse 1989). In Great Britain a review of estuaries showed that even with the highly regulated planning system there, in 1989 there were 123 cases of land enclosure affecting just over 1,500ha of the estuarine resource. This was additional to the 91,500ha lost in historical times from the major estuaries (Davidson et al. 1991). Overall these figures suggested that the current rate of loss at between 0.35% and 0.5% is no less than the historical annual rate over the last 150-200 years (0.2-0.7%). In addition to this are the less spectacular cumulative losses, of individual habitats and sites referred to in the habitat management chapters.

If the traditional approaches to conservation have failed to halt the destruction of the natural environment it is important to consider whether other methods might be more effective. Why does it appear that destruction of coastal areas continues at an increasing rate and will the adoption of more integrated coastal zone management really provide a more positive way forward? This chapter attempts to define integrated coastal management and discusses the implications of the approach for nature conservation.

13.2 Defining the coastal zone

This book has concentrated on the conservation management of individual coastal habitats, combinations of habitats (including estuaries and deltas) and their component species forming recognisable coastal systems. As intimated above, geomorphological processes are crucial to the development of coastal habitats and the wildlife they support. In this context whole catchments may provide material to the coastal margin as sediments are moved from eroding uplands to the sea. Sea level change, storms and tides provide the driving force from the sea. Looking more widely at the biological system also emphasises the links between the land and the sea, particularly where fish such as salmon move between the two. Taken together, these components help to define a zone of considerable complexity, the limits of which may be very wide (Figure 13.98).

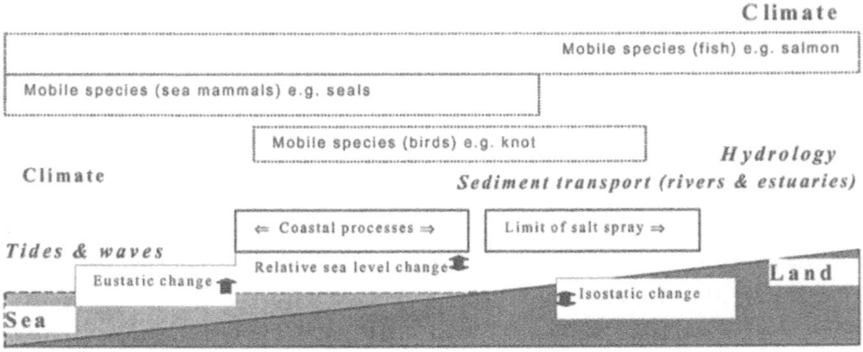

Figure 13.98. Principal components of the 'natural' coastal zone

For management purposes the zone must also be defined according to the human uses and activities which occur there. Such uses are equally wide-ranging and the scale and complexity of the biological interactions are great (Figure 13.99).

Taken from both of these perspectives, the zone is defined by a combination of natural features and human activities which may interact across the whole zone or within individual components of the zone. Recognition of this interaction is the first stage in understanding the need for an integrated approach to management of the coastal margin and in defining the way in which it should be approached.

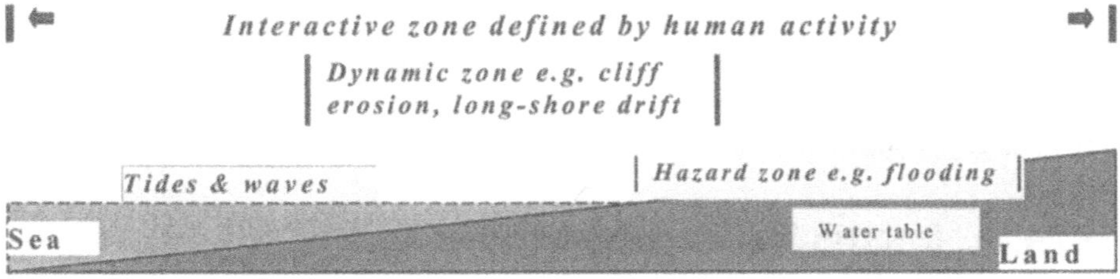

Figure 13.99. Zones for coastal management

13.3 Coastal management.

From a nature conservation point of view coastal zone management could be defined as a means of providing a framework for the development of integrated strategies for the protection of the natural coastline and marine areas, including the dynamic operation of coastal processes. This definition, however, ignores the importance of human use and exploitation of the zone and the impact on its ability to accommodate change without loss of conservation interest. Thus a more pragmatic definition might be:

Integrated coastal management provides a mechanism which facilitates the sustained use and exploitation of resources without degrading the environment, within a zone defined by natural processes and human activity.

13.3.1 The integrated approach

Those who have studied the impact of human activities in the coastal zone generally agree that a more integrated approach to management is required if sustainable development is to be achieved without further serious degradation of the natural and cultural environment. This consensus has developed over many years as academic research has been applied to an understanding of the interactions between coastal processes and human activities. In this context it is not surprising that some of the more important academic publications are from geomorphologists (Bird 1984, Carter 1988, Carter & Woodroffe 1994). The importance of understanding the dynamic nature of the coastline and the links between its component habitats and the role of human activity in changing its structure and function is central to their thinking.

The principles have been promoted by a number of organisations such as the Organisation for Economic Co-operation and Development (OECD 1993a, b), International Union for the Conservation of Nature (IUCN 1998) and the United Nations Educational, Scientific and Cultural Organisation (UNESCO 1997). The wider issues of river basin (catchment) management are also promoted through the United Nations Environment Programme (Coccossis 1997).

13.4 International / National Policy

13.4.1 The Rio Summit

In recent years much has been said about the concept of sustainable development, perhaps better expressed from a nature conservation point of view as 'sustainable use'. In many ways this embodies some of the principles which have been expressed in recent political statements. For example, Agenda 21, Chapter 15 of the report of the United Nations Conference on Environment and Development (the Earth Summit) deals with Biological Diversity and includes:

a. "conservation of biological diversity"
b. "sustainable use of biological resources" as key goals.

Chapter 17 specifically deals with the *Protection of the oceans, all kinds of seas, including enclosed and semi-enclosed seas, and coastal areas and the protection, rational use and development of their living resources*.

In Agenda 21 signatories to the Convention on Biological Diversity committed their Governments to action for the sustainable development of coastal areas and the marine environment. The growth in coastal populations, including the fact that many of the World's poorest people are concentrated on or near the coast and the recognition that its resources and environment are being 'rapidly degraded', were identified as the basis for action.

13.4.2 The United Nations Environment Programme

The United Nations had earlier identified the complexity of the environmental problems of the oceans and identified the importance of examining solutions in an integrated way. In 1972 the governing body endorsed a regional approach to the control of marine pollution and management of marine and coastal resources (UN Environmental Programme 1972). In 1973 it began to sponsor a series of Regional Seas Programmes each of which required an Action Plan. In 1990 there were 15 programmes involving 120 countries (Hinrichsen 1990).

In 1975, under the auspices of the United Nations an Intergovernmental meeting of Mediterranean coastal states approved the Mediterranean Action Plan (MAP). The main objectives are to develop integrated coastal management as a contribution to sustainable development and environmental protection for the entire Mediterranean Basin (Hinrichsen 1998). As such, they embody the principles set out at the Rio Summit and have moved much closer to them than the original work, which concentrated mainly on pollution issues.

13.4.3 The European Union

At the European level a workshop was held in Poole, Dorset in April 1991 and a communication issued which stressed the importance of the coast and the need for a more integrated approach to management in the face of its continuing degradation (Anon 1992a). This initiative was developed further at the Coastal Conservation Conference held in Scheveningen, Holland in November 1991 (Anon 1992b).

Article 2 of the EC Fifth Action Plan "TOWARDS SUSTAINABILITY" (Anon 1992c) established the framework for EU policy to the end of the 20th century; it introduces the concept of sustainable development and includes several target areas of which the coastal zone is one. **Objective: Sustainable development of coastal zones and their resources in accordance with the carrying capacity of coastal environments.**

Taking the political statements at face value, maintenance of biological diversity is closely tied up with the development of policies which embody the concept of sustainable use. In recent years there has been an upsurge in interest at both international and national levels in the concept of integrated management in the coastal zone. Some of the initiatives are outlined below.

13.4.4 A European Strategy?

Concern about Europe's coasts and seas is not new. Following his classic surveys of the British coast (Steers 1969a and 1973), Prof. J.A. Steers visited many European countries at the end of the 1960s. His report (Steers 1969b) gives an account of the state of the coast at that time and of his concerns for the future. "I must confess that my travels have left me depressed. Many, many miles of coastline are now no longer in a natural condition. Tourism, especially in the Mediterranean, has made an enormous change in the last fifteen years". He went on, "There is only one really effective way of dealing with the use of the coast – an overall state plan carried out by people who understand the economic, urban, tourist, agricultural, conservation and other demands on the coast."

Resolutions from the Council of Environment Ministers asked the European Commission to prepare a comprehensive strategy on integrated management and planning in the Community coastal zones providing a framework for its conservation and sustainable use. This was said to be in the final stages of internal discussion in December 1991. However, as late as March 1994 no such strategy had appeared when the Council of the EU adopted a resolution inviting the Commission to propose "a Community strategy for the integrated management and development of coastal zones, based on the principles of sustainability and sound ecological practice" within 6 months.

Whilst, today, a 'state plan' for the coast may be considered neither feasible nor desirable, those responsible for policy formulation and management action must be aware of the wider context in which they work. Building more effective options for future action is dependent on a better understanding of the wide range of issues and problems that beset Europe's coast and seas. In this context the European Union Demonstration Programme on Integrated Management in Coastal Zones set out to show how a more sustainable approach to the use of the coast could be achieved.

13.5 The European Union Demonstration Programme

The European Union Demonstration Programme on Integrated Management of Coastal Zones was conceived as a joint activity between the relevant Directorate General (DG), in particular DG XI (Environment), DG XIV (Fisheries) and DG XVI (Regional Policy and Cohesion), with the support of DG XII (Research), JRC (Joint Research Centre) and the European Environment Agency (EEA). This collaborative arrangement was designed to ensure that the issues of coastal zone management would be evaluated from a truly integrated viewpoint. The programme has at its heart the aim of showing how sustainable development can be achieved through co-operation and collaboration. Integration both across sectors (horizontal concertation) and at different levels of decision-making and policy formulation (vertical concertation) is a fundamental part of the programme (European Commission Services 1996).

13.5.1 Finding best practice

The Demonstration Programme was designed to test a number of hypotheses which, if validated, would be intrinsic to emerging best practice in ICZM throughout the European littoral. The hypotheses being tested are that:

- A sectoral approach to problem-solving is inimical to securing successful outcomes.
- There are significant barriers to the flow of information between the scientific and technical communities and decision-makers in local and regional authorities, and the private sector.
- The lack of horizontal and vertical concertation frustrates a consensual approach to coastal management.
- Monitoring and evaluation of coastal resources are rarely undertaken in any consistent or robust way.
- European Community policies and actions, and those of Member States, can be contradictory or competitive in the coastal zone.
- Less than optimal outcomes because of the lack of genuine participation.
- Statutory and regulatory mechanisms are sometimes unhelpful to the integrated approach.

13.5.2 The thematic studies

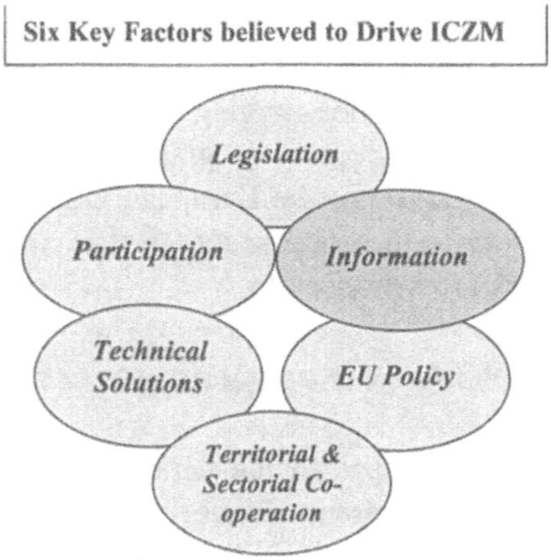

Figure 13.100. Thematic studies in European Union Demonstration Programme

The overall purpose of the Demonstration Programme was to identify what works in ICZM. The programme has three components:

1. 35 local Demonstration Projects that were chosen to show different scales, problems and type of coastal experience;
2. six thematic studies to review what works and why (Figure 13.100); and
3. a wide-ranging debate on the results of the Demonstration Programme.

The thematic studies were set up to review how integrated management can contribute to sustainable development along the coastline of Europe. The factors thought to drive this process formed the basis for the six cross-cutting studies.

13.5.3 Conclusions from the programme

The Demonstration Programme projects provide examples of good, and bad, practice in integrated coastal management (ICM) in a range of socio-economic, cultural, administrative and physical conditions. From these examples, it has become clear that, for sustainable management at the coast to be effective, it must:

- take a wide-ranging perspective;
- be based on an understanding of the specific conditions in the area of interest;
- work with natural processes;
- ensure that decisions taken today do not foreclose options for the future.
- use participatory planning to develop consensus;
- ensure the support and involvement of all relevant bodies;
- use a combination of instruments (European Commission 1999a).

Given the strategic importance of the coastal zones to the future economic development of Europe, the treaty obligations for cohesion and for integration of environment into other policies, and the major role that existing EU policy already has in shaping the coastal zones, it is concluded that there is a need for a European Strategy on ICZM (European Commission 1999a, b). These documents reinforce the need not only to take a more wide-ranging view of coastal management issues, but also to provide much greater integration between European Union policies. Given the structure of the European Union, these policies will be largely implemented through nation states. In this context it is important to look at the way these currently operate.

13.6 European governmental approaches

In Europe a wide variety of approaches have been adopted for the use of coastal areas. These range from policies specifically protecting the coast, which may include the definition of a legally protected strip along the coast, to general policy statements extolling the virtues of adopting a more integrated approach based on existing legislation.

13.6.1 Coast specific, prescriptive policies

A number of European governments have specific coastal laws, for example, in Spain the Coastal Law (1988) and in Portugal the Law on Integrated Management of the Coast (1993). There is often, associated with this, designation of a strip of land from the 'edge of the sea' defined as high water mark or some other suitable reference point, which is subject to control of development. Examples are listed in Table 13.32 below taken from Nordberg (1995, 1999).

Table 13.32. Statutory distances for coastal protection for some countries in Europe

Country	Type of Protection	Distance (m)	Definition
Finland	Planning controls	100-200	From the coast to the limits of influence by the sea
Denmark	Except in urban localities where the 100m line applies - no development	100-300	From the position above the shore where continuous vegetation begins
France	General development prohibition within 100m, extended to 200m in especially sensitive areas	100-200	From the high tide mark
Spain	Protection from development which can be extended to 200m if local governments agree	100-200	To the limit of the shore landwards
Portugal	Mandatory terrestrial protection zone where major developments require a land-use plan	500	Within a 2km zone from the shore
Italy	No development without an approved landscape plan	300	From the high water mark
Turkey	Non-development zone	100	

Most of these examples also include a further protected zone, up to 3km in the case of Denmark, where further restrictions apply. This may involve the control of specific types of activity such as in France, where there is a general prohibition of roads within 2km or more of the coast. Usually development is allowed only in the context of an agreed land-use plan, as in Portugal.

13.6.2 Use existing policies

The second approach is only 'coast specific' in the sense that existing, usually land-based, planning systems, are modified to cover the special needs of the coastal zone. In the UK and Holland there are no general provisions for the protection of the coast. The comprehensive planning legislation, which is implemented in the UK by local planning authorities, has to take account of national policy guidance. The Policy Guidelines for the Coast prepared by the Department of the Environment in 1995 specifically state that "Development which does not require a coastal location should normally be placed inland.".

In Holland all natural and semi-natural habitats in the coastal zone (sand dunes, salt marshes, mudflats, estuaries and shallow parts of the North Sea) have been given the principal function of nature conservation since 1979. This has been implemented in physical planning schemes at provincial and local level and has resulted in it having a special status where it is protected against sectoral developments. The same coastal habitats were all included in the national ecological network, which was approved by the Dutch government in 1990 (Ministry of Agriculture, Nature management and Fisheries 1990). Since 1990 in Holland, where coastal protection is an important issue, policies have been enacted which include the concept of "dynamic preservation". These set out to maintain the coast in its present location by counteracting erosion using methods such as beach-feeding. Within this context natural dynamics are allowed to take their part in protecting the land from flooding (de Ruig 1998). Two different models are described by Carter (1988) which he called the "British" and the "American" models. An example from the USA is described below.

13.6.3 An example from the USA

The Commonwealth of Massachusetts provides an example of the way in which the Coastal Zone Management Act, which Congress passed in 1972, has been implemented in the USA. In 1978 the Massachusetts Coastal Zone Management Programme received approval from the Federal Government. With this approval came grants of $16.7 million up to 1991. Within this legislative framework the Massachusetts office in Boston (with 25 staff) oversees the programme on behalf of the State. Whilst this is mainly a planning and policy agency it does review and comment on proposals for coastal development. Overall the programme seeks to promote, amongst other things:

- Protection and enhancement of existing coastal wetlands and the creation of new ones;
- Preventing or significantly reducing risk to human property and life by eliminating development in hazardous areas;
- Increasing opportunities for public access;
- Control of cumulative impacts of coastal development and use of resources;
- Preparing and implementing management plans for special conservation areas.

The cascade of approaches involves mechanisms for funding co-ordination and implementing management action at each level and is highly structured and based on a programme agreed and grant-aided by the Federal Government. Some of the reasons for adopting this approach include recognition that at local level there may not be sufficient capacity in terms of "financial resources, technical ability or political will" (Beatley et al. 1994). Added to this is the fact that coastal (and marine) problems often cross territorial boundaries. The financial help from the Federal Government is a key part of the development and implementation of comprehensive management plans embracing all the issues affecting the use of the coastal zone and its environmental conservation.

13.7 Other institutional methods

It is clear that, on land, a range of measures are used to protect and manage coastal areas. These include prescriptive, often site-based, legislation to more flexible institutional and administrative arrangements which seek to encourage co-operation between sectoral interests. Generally the protection of sites and species of value from a nature conservation point of view continues to be the subject of prescriptive site-based legislation such as the highly significant European Union Habitats and Species Directive. In recent years, however, there has been a shift from policies which define relatively narrow nature conservation interests within specified boundaries, to more broadly-based approaches which reflect the complexity of whole ecosystems (such as estuaries and deltas dealt with in Chapter 10) and the human use to which they are put.

Whatever method is adopted it is equally clear that effective conservation can be achieved only when there is a willingness to enforce the legislation or enact the appropriate management response. This may be very difficult where economic or social interests are seen to be threatened. In this context it is often the way in which the policy is put into effect (the process) which is as important, if not more important, than the actual policy decided upon. The general approaches to identifying issues, developing policy and devising management plans are described next.

13.7.1 Regional seas programmes

The question was posed at the beginning of the previous chapter as to whether the wider approach to CZM had helped the cause of nature conservation. The case is not yet proven, though there are some encouraging signs. At an international level, inter-governmental initiatives were enacted for the Mediterranean and between member states bordering the North Sea.

In the Mediterranean an Action Plan was produced as part of the UN Regional Seas Programme. Here the early concerns centred around the issue of coastal pollution and the Barcelona Convention for the Protection of the Mediterranean Sea against Pollution became the main legal document. Three of the protocols were:

- the prevention of pollution at sea by dumping (1978);
- combating pollution in an emergency (1978);
- prevention of pollution from land-based sources (1983).

Later conservation issues were more specifically addressed when a protocol on Special Protection Areas (1986) was enacted. Throughout this period it was recognised that 80% of pollution came from land-based sources and the new Coastal Areas Management Programmes (CAMP) are extending their concern to other key issues. These include a wider range of subjects (14 in the case of the island of Rhodes) where tourism, over-exploitation of natural resources, environmental degradation and fragmentation of institutional structures for management were amongst the issues identified (Leftic 1995).

In the North Sea the issues considered in the first Quality Status Report, 1987 provided an overall review of the impact of human activities on the North Sea. In 1990 at the third international Ministerial conference the 1990 "Interim Report on the Quality Status of the North Sea" drew attention to other issues of concern, notably algal blooms and viral decease in seals, and provided an update on the input of contaminants and their effects. In 1993 species and habitats of the coastal margin were added to the areas of concern in the North Sea Quality Status Report, 1993 (North Sea Task Force 1993). This was accompanied by a description of the North Sea coastal margin of Great Britain (Doody et al. 1993) which include species concentrations, habitats, fisheries and geology amongst a wide variety of information covering human use of the whole of the coastal area.

13.7.2 Inter-departmental co-ordination

The delay in the production of a EU coastal strategy is thought to have been due to conflict between the Directorates responsible for conservation and regional development, the latter considering that coastal management would inevitably put a break on development in less favoured areas of the Union. The fact that there is greater co-operation through the European Union Demonstration Programme on ICZM may, in part, be a recognition that it is in all our interests to manage the coast on a sustainable basis. It may also be that it is recognised that the past reluctance to take account of the environmental consequences has not always been in the best interests of the developments themselves. The case of the change in attitude to tourist urbanisation on the Costa del Sol and in the Balearic islands, Spain is an example. Here declining tourist numbers was associated, in part, with a deteriorating quality of environment. This has been rectified in Majorca and Ibiza with a much tighter control over the scale, location and type of development and a general reaction against further major tourist development by the local population.

Given that the Mediterranean is one of the world's most popular holiday destinations with 25% of its hotel accommodation (Cori 1999), it is essential that a more integrated approach between departments is developed. Without it the coast, here at least, may be doomed to be choked by people, their tourist infrastructure and cars, 99 million of which were registered in 1984 (Cori 1999). This intense use inevitably leads to conflict. The problem is to manage the dynamic interrelations between the three key elements which make up the coastal system: the geomorphology, ecology and economy (Green & Penning-Rowsell 1999). This is clearly not easy and whilst we know much about the first two in terms of science we know little about the way the third can be integrated with them.

13.7.3 The emerging eastern European states

The results of the democratisation of the former Soviet Union are just beginning to emerge. The situation on the islets off the coast of Estonia will serve as an illustration of the implications for coastal management and conservation. Here in the western part of the coast the islets are rapidly evolving, a process intimately bound up with human use. Agricultural activities have prevented forest covering the land and a diverse landscape of hay meadows, cultivated fields (on some of the larger sites) and small woodlands has developed. Grazing also prevented the growth of scrub in the higher land and dense reeds on the shore. By the 1930s despite many 'agricultural improvements' and changes in the landscape, the land-use had been "on the whole, environmentally sound" (Puurmann & Ratas 1995, Ratas & Puurmann 1995). Later a variety of changes has resulted in the abandonment of much traditional farming activity. As farms were amalgamated, overgrazing took place and the cultural landscapes and their wildlife compromised. Abandonment occurred both because of economic pressures, and ironically, mistaken management in some areas where land was taken into protection for nature conservation.

It is clear that the restoration of these cultural landscapes and their wildlife is intimately bound up with the political system which determines the form of land ownership and allows more traditional forms of management. It is important to recognise that as the new states are brought into the European Union the drive towards more efficient and intensive agriculture respects the existence of a more traditional approach which sustains both local populations and high quality wildlife.

13.8 Local initiatives; raising awareness and promoting understanding, the role of research

There is one fundamental requirement in promoting a more sustainable approach to resource utilisation in the coastal zone, especially at a local level. This involves the provision of information and its deployment in such a way as to ensure effective action is taken. Often a considerable amount of scientific data is available which points up key elements of understanding. In the United Kingdom a major research study, the Land-Ocean Interaction Study (Natural Environment Research Council 1998), and the wider-ranging Land-Ocean Interactions in the Coastal Zone (LOICZ), a core project of the International Geosphere Programme, both point to the need to look across the land-sea interface when considering coastal management. The models derived from these studies are often detailed, elegant and may include important indicators for evaluating economic consequences of policy responses (Turner et al. 1998).

The link between management in the catchment and the effect this has on the coast has been described in this book, especially Chapter 9 and in the definitions of the coastal zone (Figures 13.97 & 13.98 above). It is also used to expose the way in which damming rivers can create conditions whereby erosion takes place along the coast (Bird 1996). Acting upon the understanding of this in terms of policy formulation requires an acceptance on the part of the sectors and the population affected that the link is real. However, understanding that the loss of a beach and flooding of the recreation area from the sea (Figure 13.101) could be due to actions taking place in the hinterland may be difficult. One key issue lies in translating the scientific knowledge into understanding. Four case studies in the United States pointed up 5 key factors in making scientific information more influential in determining policy. Amongst these there needed to be "clear evidence of change"... "consensus among the scientists and identification of effective and feasible solutions to the problems" (Boesch 1996). All too often the interpretation and communication of the results to a wider audience are not considered until late in the process of research, if at all. As the case histories in the US show, much better integration is required not just between scientists, but in communicating their understanding to those who take the decisions. In this case, especially with the "engineering communities working on coastal ecosystems, watershed processes, agricultural practices and waste treatment" (Boesch 1996).

Figure 13.101. Damming rivers for irrigation in Morocco to improve agriculture has reduced sediment supply to the beach at Agadir, with serious implications for the tourist industry

13.9 Implementing Integrated Coastal Management

Implementing effective coastal management to achieve sustainable development and protect the environment is not easy. The combination of stakeholders and the complexity of the issues can be enormous. Attempting to deal with all the possible interactions through institutional mechanisms can result in highly complex procedures. The Demonstration Programme project for Dorset, in southern England, for example, produced 15 topic papers. 450 responses, a 'Forum steering group' and three working groups were used to prepare the draft strategy. From these discussions 9 priority areas were identified upon which the strategy was based (Dorset Coast Forum 1999).

Whilst this level of consultation may not be appropriate in all cases, a key feature of all successful local initiatives lies in agreeing real issues and acting upon them. In this context the stakeholders must have confidence that their interests are being taken into account and that the solutions address the issues. It is the process through which this confidence is built which is so important to the development, and implementation of sound policies and management action.

14. HABITAT PROTECTION, COASTAL CONSERVATION & MANAGEMENT

14.1 Introduction

This final chapter examines the successes and failures in protecting coastal habitats and sites. It reviews the role of change as a natural component of coastal systems and examines the way in which this is viewed by the conservationist, the developer, and those concerned with sea defence and coast protection.

14.1.1 Site protection and management

Previous chapters have been concerned with the development of both traditional and innovative approaches to nature conservation management for all the main coastal habitats. It is clear that despite the extensive destruction of many coastal areas there remain substantial, and possibly unique, nature conservation sites throughout Europe and elsewhere in the world. A first requirement for promoting coastal conservation is to prevent further losses of undisturbed habitat. A second is to manage the remaining areas of interest so as to maximise the chances of survival of the 'natural' components of plant and animal communities present at a particular location.

Site management for nature conservation is, in theory, relatively easy. Once a site is identified as important and in need of protection, then wildlife management techniques are mostly tried and tested. However, principles which recognise the interrelated nature of the natural environment (habitats and species) and of human use are less completely understood. A key driving force for the development of coastal systems lies in the geomorphological processes which operate there. Often the existence of one habitat is entirely dependent on another (a shingle bar may form the fabric upon which a sand dune develops and in its turn this may allow saltmarsh growth). The coast is also often highly dynamic and change is an important part of the development of both individual habitats and combinations of habitats.

In one sense the definition and establishment of nature reserves are an anathema to conservation (or wise use) of the coastal zone. By necessity a nature reserve has to be delimited, both by a line on a map and often on the ground by a fence or similar physical feature. Often only parts, and sometimes only small parts, of individual sites can be acquired and suitably managed. This has also led to a preoccupation with the protection of existing interests, identified when the site was first assessed. This can mean that management across the whole geomorphological system may be impossible. Thus the natural ability of the coast to accommodate perturbations in the environment, whether caused through human use or natural forces (tides, storms, sea level etc.), are often not taken into account. Superimposed on this is the impact of human use which has in many areas had a profound effect on the nature of the 'natural' coastal land form.

14.2 The coastal 'squeeze' - a new paradigm?

Loss of habitats through the exploitation of the coastal zone for human use and occupation have been emphasised throughout this book. Historically, the position of sea level has moved relative to the land depending on the interplay of isostatic and eustatic changes. In low-lying coastal areas such as the Wash (Shennan 1989) and the Severn (Allen 1992), for example, this resulted in periods of landward or seaward movement of the shoreline depending on the balance of sedimentary over erosive forces. The process of land enclosure results in a replacement of a sometimes soft and flexible barrier at the edge of the sea with a straight, hard inflexible artificial structure. These barriers restrict the ability of the coastline, to provide a flexible response to changing patterns of sea level rise and storms. In a zone which is recognised as being subject to the forces of nature, it could be argued that humankind has devoted too much effort to the 'battle with the sea'. Given this scenario it is right to consider whether any lessons can be learned from the natural world, which might allow a more enlightened approach to the management of the coast.

14.2.1 Changing perceptions: embankment in the southern North Sea

The saltmarshes around the southern North Sea have been progressively embanked and converted to some of the most productive agricultural land. In the Wash, Norfolk and Lincolnshire, it has been recognised that this process destroys the saltmarsh habitat in the short term and reduces its overall diversity by continually removing the more mature and botanically diverse sections of marsh. Although new marsh develops outside the sea bank this does not fully compensate for the losses. Enclosures between 1971 and 1985 resulted in the loss of 865ha of marsh. Accretion rates of over 20mm per annum were recorded in front of these enclosures, when new marsh totalling 781ha developed over the same period (Hill 1988).

Although the evidence is not conclusive, Evans & Collins (1987) suggest that low water mark is not advancing progressively seawards. Thus as the saltmarsh regenerates outside the sea-wall, there appears to be a consequent loss of sand and mud flats in the Wash which provide important winter feeding areas for large numbers of wildfowl and waders. This scenario is supported by evidence from a Sea Defence Management study prepared for the National Rivers Authority in 1991 which concluded that 70% of the coastline from Humberside to the Thames was retreating.

On the basis of this evidence further piecemeal enclosure was resisted by conservation bodies at a public inquiry into saltmarsh enclosure at Gedney Drove End in 1981. A local inquiry into a moratorium on enclosure proposed by the Lincolnshire County Council in their coastal subject plan in 1983 was supported. Happily no further enclosures have taken place since 1979. This example highlights three important consequences of saltmarsh enclosure and agricultural reclamation, namely:

1. Losses usually involve more mature high level plant communities and their associated invertebrates and breeding birds;
2. Losses reduce the available feeding for winter grazing for ducks and geese;
3. Reduction of the intertidal sand and mud flats where new marsh, in some cases with an accelerated accretion rate, extends beyond the new sea wall.

The process including subsequent conversion of these areas to intensive agriculture reduces the overall nature conservation value as saltmarshes and grazing marshes are destroyed (Chapters 5 & 6). Recognition of this 'squeezing' of the intertidal land together with studies on the loss of saltmarsh on the Essex Coast further south (Burd 1992) were partly responsible for a change in attitude towards protection of the coastline 'at all costs' and helped establish the programme of experimental 'managed retreat' discussed in Chapters 5, and 12.

Similar arguments from conservationists took place in Germany where enclosure of tidal marshes continued up to the 1990s. Since the 1960s the primary objective of embankment has been for sea defences. However these have continued to have a detrimental impact, especially on the rich areas of the German Wadden Sea. A decline in projects to enclose marshes did not take place until a number of factors came together to persuade the authorities that continuing enclosure was not in the best interest on the environment or for sea defence. These were:

- objections from environmentalists;
- a reduced need for greater agricultural production;
- new more environmentally sound legislation;
- recognition that sea level rise was an important issue (Goeldner 1999).

Since 1982 approximately 2,000ha have been "reintegrated with the sea". The embankments for the first two polders to be treated in this way (1982 and 1984) were maintained so full 'reclamation' was not possible and to some extent the policy was designed to assuage the environmentalists whilst continuing the policy of enclosure. However the more recent "reintegration" (1988-1989) was much more a planning requirement to compensate for losses through embankment elsewhere (Goeldner 1999). It remains to be seen whether the trend towards abandonment of land to the sea becomes more acceptable as a more effective means of sea defence especially in areas of rising sea level as in these two examples. The evidence from elsewhere in Europe suggests that they will continue to be an exception rather than accepted practice.

14.3 Sea level change

An important question for any policy maker or manager dealing with low-lying or dynamic coastal areas is the likely impact of flooding and/or erosion. Sea level change is a major factor determining the long term management strategy for dealing with this issue. As has already been described, intertidal land has been removed from the influence of the sea through enclosure. In areas where sea level is rising as a result of isostatic adjustment (change in level of the land due to removal of the glacial ice sheets), there is a further 'squeeze' resulting from a steepening beach profile and foreshortening of the seaward zones. A predicted sea level rise, accelerated by global warming will cause a further 'squeezing' of the natural tidal land. In the face of this, building bigger and 'better' sea defence structures may not be the answer, particularly where these are being undermined by falling beach levels.

On a world scale shorelines have been reported as showing a prevalence of erosion over accretion and sea level rise has been implicated (Bird 1985). The combination of continuing habitat destruction and sea level rise has important consequences for the survival of the remaining coastal areas of nature conservation interest as evidenced by the loss of saltmarshes in southeast England.

14.3.1 A shared problem

Rather than continuing the battle with the sea, there is an argument which suggests that working with nature rather than against it may have long term benefits. This may include reversal of the present preoccupation with land development and coast protection, to more concern for widening the coastal zone in order to take advantage of natural processes both in the interests of wildlife conservation and coastal defence. A powerful alliance can be built up when common goals are agreed. An example of this can be found in the experimental set back (managed retreat) options being developed by English Nature and the Ministry of Agriculture Fisheries and Food in the UK (Burd 1995).

As has been stated already, far from being static or progressing in an orderly fashion from one successional stage to another, coastal habitat development is complex and change is an integral part of that complexity. Accepting this may be equally important to both the conservation manager and to those concerned with coastal defence or flood protection. It may be no accident that during a severe storm in 1991 caravans west of the Point of Ayre on the Dee Estuary (north Wales) were not affected, partly because the volume of sand in the dunes replenished the beach and absorbed the impact of wave energy. By contrast the village of Towyn, only a few miles to the west was badly flooded, when the man-made sea defences failed.

If we take our cue from nature and develop a better understanding of the way in which the coast operates it should be possible to develop strategies for its sustainable use both for the benefit of wildlife and man. This will also require a better understanding of human influences on the coast and a more enlightened approach to its management. The challenge is to achieve the most appropriate level of interference which makes maximum use of the natural ability of the coastline and coastal habitats to adapt to change.

14.4 Changing attitudes - dynamic conservation?

The example of sand dunes described in Chapter 8 shows how a better understanding of the way in which coastal habitats behave may change our attitude to their conservation management. Far from being fragile areas they are in fact quite robust and in the early stages of development, at least, are able to accommodate changes in sea level, storms and other perturbations including human disturbance. The dynamic nature of the habitat is one of its strengths and attractions (Figure 14.102). Saltmarshes are also very dynamic and include sequences of erosion followed by regrowth as estuary channels change their course, and are very responsive to sea level change.

Figure 14.102. Råbjerg Mile, Skagen Odde, an eroding dune, provides a 'natural' attraction on the north Danish coast

Sea cliffs composed of soft rocks rely, in many instances, on mobility as an essential part of the process of retaining the biological nature conservation interest (Chapter 3). Thus it can be seen that far from being static or progressing in an orderly fashion from one successional stage to another, coastal habitat development is much more complex and change is an integral part of that complexity.

The nature conservationist may readily embrace concepts of change, as part of the normal course of habitat development, particularly in the early stages of succession. The prevention of change, in the sense of arresting succession in older more stable habitats, may also be acceptable where this is seen to be detrimental to the nature conservation interest of a particular site. It has been less easy to consider change as a positive force in conservation management.

Typically, erosion of sand dunes will elicit a desire to 'protect' them and considerable energy has often been expended on building sand fences and controlling people. Continuous disturbance around car parks and access points can result in the destruction of vegetation. However, if the system is given time to heal and the cause of damage removed, vegetation can become re-established without further interference. Some forms of physical damage can cause a change which becomes a permanent feature, such as tracks left by vehicles on mature shingle. However, these should be separated from the more natural forms of change which may be an essential component of the proper development of the system.

If we take our cue from nature the management options may be diversified. Thus, the conservation manager may consider a more adventurous approach involving positive management for change though this may not be easy. Taking a more radical solution may be difficult to reconcile with traditional approaches, especially if the actions are constrained by management objectives based on a limited understanding of the dynamics of coastal systems.

The protection of the remaining areas of natural and semi-natural habitats will continue to be a primary nature conservation aim. For this to be successful the overriding requirement will be, as it is now, to prevent further direct damage and destruction through human activities and manage the areas appropriately. However, this can only be part of the solution. The operation of natural processes which sustain the wildlife importance of many areas has already been compromised by habitat destruction and management which has allowed over-stabilisation to occur. In this context individual habitats, such as sand dunes, may need to be managed in a more dynamic way to ensure the development of the full range of dune types (Doody 1989).

In response to this changing paradigm the Dutch Government, in 1990, made a policy decision to stop any further loss of its coastline. This attempted to reconcile the twin demands of protecting nature and the safety of the population from flooding through 'dynamic preservation'. After 5 years this policy option was assessed and it was concluded that beach nourishment was effective in maintaining the sea defence function as well as the nature conservation value of beaches and dunes. However, a doubling of the sediment supply to the beach will be needed and it is not clear where this will come from if sea levels continue to rise. An even more 'integrated' approach will be required if a sustainable balance between socio-economic development and maintaining the dynamic system is to be achieved (de Ruig & Hillen 1997).

14.5 Integrating management

The development of more integrated approaches to coastal management has been achieved through the preparation of a 'Management Plan'. This is usually undertaken in consultation with individuals or groups of individuals identified as having a direct interest in the features of importance on the site, concerned with its management or exploitation. Whilst this provides an important stage in data gathering it fails to involve them directly at an early stage in the process, including introducing the concept of integrated management. Involving people more closely in the development of the mechanisms for gathering, analysing and reporting on the needs of a particular area provides more ownership of the outcome of any review than is achieved by the preparation of an 'off the shelf' plan.

It is recommended therefore that a wide-ranging approach to the development of a coastal zone policy is undertaken. This must identify, at an early stage, the key players who should be involved. Contributions should also be sought from those sectors whose activities have been shown to affect the area as well as local and national authorities, non-governmental organisations and the public.

For those concerned with nature conservation management, this will also require a more enlightened approach to be adopted. This must embrace the principles of coastal zone management, including acceptance that geomorphological processes are important to the conservation of individual habitats.

14.5.1 Local approaches

Local approaches to integrating policies in the coastal zone have been developed in many areas by local authorities as well as through national and international bodies such as the United Nations. It seems surprising therefore that many European governments seem to be hesitant about implementing anything other than sectoral approaches to management or the production of plans for individual sites and stretches of coast. The reason for this may lie partly in the way in which the issues are portrayed by those promoting a more integrated and holistic approach to the coast and marine areas. The complexity of the interactions of both the natural processes and human activities are highlighted. At the same time the conservation sector, by stressing the scale of loss of coastal habitats and general degradation of the environment, reinforces the view that its protection can best be achieved by some form of legislation. Perhaps this leads to the perception by those responsible for formulating national and international policy that a prescriptive, complex approach backed by legislation is all that is being advocated.

14.5.2 Sustainable use

It is unrealistic to believe that on the coastline of Europe, or anywhere else around the world, all but a very few specially protected nature reserves or inaccessible areas will be free from the destructive activities of humankind. Therefore its conservation will rely on the integration of human activities, including economic uses, in a way that can be sustained with minimal damage to the environment. This remains the greatest challenge and for some the true expression of the aspirations of coastal zone management. This will not only require a knowledge of the resource but also the scale and impact of human activities on it. Armed with appropriate information and understanding, it should be possible to devise strategies which accommodate the needs of wildlife and human use. This will require mechanisms to be developed which bring together all those who influence action in these areas, not only the sectional interests as at present.

Given the extent of devastation and destruction of the coastline, it could be argued that we have already reduced its natural components to an unacceptable level through the process of land claim for industry, housing roads and the like, and for agriculture. Coupled with this, the preoccupation in many areas with the maintenance of what has become a static line of defence against the natural forces of the sea, further compromises the natural functioning of the zone. If the sea level rise predictions are correct this zone may be further depleted as erosion becomes more prevalent. Under these circumstances restoration of some of the natural processes and recognition of nature conservation as an equal partner in developing strategies for the protection of land and property will be essential if we are also to protect wildlife.

This could be taken even further with recognition that natural coastal systems may in some cases provide a more efficient barrier to sea level rise and storms than conventional hard defences. It might also lead to the acceptance that some of the problems of coastal defence and instability are the product of inappropriate use of, and interference with, coastal areas and processes by humankind. The natural coastline might then be recognised as a valuable asset in its own right; critical to the sustainable use of the coast, rather than something requiring control and strengthening in the 'battle' with the sea.

14.6 Participation and consultation

One of the conclusions from the European Union Demonstration Programme on Integrated Coastal Zone Management was that using participatory planning helps to develop consensus. The projects provided examples of the degree of consultation and the methodology used. Two basic models were elucidated (Doody et al. 1999):

1. The participatory model, where all the stakeholders together identify the issues and information needs as part of the process of developing more integrated approaches to management. There is no predetermined structure;

2. The consultative model, where the players (regulators, usually planning authorities, developers and specialists who advise the regulators) prepare a plan for consultation. The plan, therefore, has a predetermined structure and embodies assumptions as to what issues are relevant. Consultation, such as it is, takes place amongst the regulators. In this model the wider public are restricted in their ability to input comment.

Assuming that sufficient care is taken, either a *participatory* or a *consultative* approach can work. However, there are distinct advantages in adopting the first. These include generating a sense of ownership of the issues and policy options, helping to establishing a common perception of the resources within the area and the human activities that impinge upon them and understanding the constraints on policy implementation. Often one of most common shared problems on the coast is the loss of land to erosion (Figure 14.103)

Figure 14.103. Debate on what to do on the east Anglia coast of England - protect or set back

14.7 The future

Nature conservation and management are intimately bound together. Coastal management in particular must operate at a variety of scales if it is to ensure the coast and its resources are used in a sustainable way. This book has tried to provide a sequence of discussions describing the conservation value of the coast and the way in which human activities have influenced it. It has taken the reader from the management of individual habitats to more integrated approaches. This has included consideration of ecological networks of sites and species.

Traditional conservation on the coast, as elsewhere, has in the past been primarily concerned with the protection and management of habitats and species concentrations deemed to be of special importance for their flora, fauna or geological interest. In recent years this relatively narrow approach has broadened to a recognition of the value of promoting more integrated forms of management. With this comes the need to have a wide knowledge of the coastal and the marine environments, the hinterland and its influences. In addition, it is essential to know and understand the economic and social influences on the coast and how they interact with it. In the past these two aspects have rarely been brought together. When they have, it has often been in conflict, with one party (the environmentalist) opposing the other (the developer) in their quest for new or enhanced opportunities for exploitation.

A necessary first step in accepting a shared responsibility for sustainable economic development which conserves the coastal environment is, therefore, a common understanding of the key issues. Given the long history of use and the enormous pressure which the coasts and seas of Europe and elsewhere continue to absorb, it is a reflection of their resilience that so much of interest still survives. In this context two principal components are needed for the delivery of the sustainable development of the coast: recognition of its landscape value and the importance of the legacy of past human activity which has helped shape its present character. The UN Agenda 21 reminds us of the dual responsibility to the environment and the need for sustainable use of the coast and its resources.

14.7.1 The politics of sustainability

Without this wider understanding, it is suggested that the politics of sustainability rest more on human perceptions and values than on the intrinsic worth of 'natural' systems. Therefore, if a fisherman can be convinced that conservation interests overlap with the need for nursery areas for commercial fish stocks, or the engineer that saltmarshes are good for coastal defence, then it will be possible to build more effective partnerships to manage the coast. Achieving this may be difficult especially where economic forces drive the needs of people and the management of the coast. Hitherto there has been relatively little integration between the 'environmentalist' and the 'developer'.

It is becoming increasingly apparent to those who will listen that more enlightened and environmentally sound decisions will be of long term benefit, both environmentally and economically. In the Ebro Delta in Spain, for example, erosion is taking place at the margins as the supplies of sediment needed to keep the delta growing in the face of rising sea level are reduced due to damming of the rivers. That these anthropogenically-induced changes occur faster than the delta can adjust poses important economic questions for the rice cultivation which is the major activity in the delta (Palanques & Guillén 1998). The initial response has been to promote hard engineering solutions to prevent erosion. This will do nothing to counteract the cause of the erosion and measures recommended by Ibàñez et al. (1997), involving the partial removal of sediment collected in dams, will ultimately be required. Whether the local people will accept this as the best way forward is still in doubt.

14.7.2 The stakeholder principle

There is an increasing awareness by those who live, work or enjoy the coast that they have a stake in the development and implementation of policies which affect it. In this context the European Commission talks about "actors" and in the UK the Department of Environment, in its Policy Guidelines for the coast suggest "improved links between the different sectors". English Nature's "Campaign for a Living Coast" advocates "partnerships" and a Scottish Natural Heritage discussion paper on the coast outlines "initiatives based on partnerships, co-operation and local management". There is also a developing view that national or international policies are implemented through local action. The mechanism for this involves institutional and administrative frameworks which are moving towards a common approach based on integrated coastal zone management (ICZM), also called coastal zone management (CZM) or integrated coastal management (ICM) discussed in Chapter 13. The approach rests on two key principles – the need to integrate actions across competing sectors and the importance of consensus building in the development and implementation of policy.

The 'road to this more enlightened approach' will be most successfully achieved when the stakeholders are involved in the process by which the policy decisions are taken. In summary a sequence of actions might involve several interrelated activities (Table 14.33).

Table 14.33. A 'road' to more enlightened coastal management

1. Information and its collection and collation as a TOOL for involvement	Provides a neutral basis for identification and consideration if issues, including conflict resolution
2. Coastal Zone Management is a METHOD for institutional involvement and conflict resolution	Development of partnerships for policy review and implementation e.g. preparation of management plans
3. Local politics as a MEANS for developing partnerships	Use of Agenda 21 as a catalyst for the development of a more sustainable approach to development and use of the coast, allied to the maintenance of Biodiversity

This process is as important for decisions affecting sand dunes as it is for the implementation of a major coastal zone plan. Local consensus is essential to the development and implementation of policy, as has been so amply demonstrated by some of the local projects included in the European Union Demonstration Programme on Integrated Coastal Zone Management. We are, however, far from the situation where the removal of forestry plantations, originally planted to halt erosion, is seen as acceptable and a positive benefit for both sea defence and conservation. Promoting more enlightened approaches to coastal management, as advocated here, is also more difficult when they are dictated by national or international legislation. It is important to recognise that a 'one off' consultation will not be effective in maintaining the acceptance of any new approaches. Local participation must be part of an ongoing process through which management decisions and implementation of policy are 'informed' by good information.

Managing the coast is a complex task, especially when dealing with nature conservation issues. These have too often in the past been considered the preserve of the specialist or the committed environmentalist. Both have been branded as 'zealots' not versed in the real world where economic requirements dictate the need to pursue policies which may impact on the environment. I hope this book has in small measure helped to remove some of these prejudices and show that in the end we all derive a large proportion of our health both economically and personally from a sustainable coastal environment. Unravelling some of the issues and providing pointers to appropriate management for nature is only one part of the story!

Allied to this is the need to convince the general public who elect the policy makers, and ultimately determine the direction of management action that a wider perspective is needed. No one disputes the fact that the Bay of Naples will suffer an irruption of Vesuvius at some time in the future. Scientific studies and monitoring of the volcano's activity suggest this is likely to be sooner rather than later. Such an eruption could be as devastating to the local population as the one which destroyed Pompeii in 79BC. Despite this, both legal and illegal houses continue to be built on the side of the mountain. Providing timely, relevant and accessible information is only part of the story (Figure 14.104). To be of value those who receive it, they must be prepared to listen, take note, learn and act in accordance with the information provided and the knowledge gained.

Figure 14.104. Sound coastal conservation and management require a clear understanding of the environmental and human factors influencing development. 'Specialists' communicating 'knowledge', Skaw Spit, northern Denmark

In the end the coastal zone presents us with a problem only because we have chosen to live in increasing numbers along its shores. In a series of publications covering the coast of the USA (Pilkey & Neal 1984), the authors highlight a number of "Truths of the shoreline". Amongst these is the notion that "There is no erosion problem until a structure is built on a shoreline." "Whether the beach is growing or shrinking does not concern the visiting swimmer, surfer, hiker, or fisherman. It is when man builds 'permanent' structures in this zone of change that a problem develops."

In the past conserving the coast and its wildlife has been seen from a protectionist standpoint, often conflicting with other uses. If we accept human activities are an integral part of the system then coastal conservation and human exploitation have more in common than at first might appear. From both perspectives a wider appreciation of the importance of change and the dynamic nature of the coast will result in policies and management which protect the environment for wildlife and humankind. For this to be achieved we might see a wider more 'natural' coastal area as one which provides wildlife habitat, coastal protection and is an economic resource in its own right.

14.8 "Win some, lose some"

This book has been concerned with the development of more sustainable approaches to coastal management in the interests of nature conservation. In the past the drive for economic development has led to clashes between 'environmentalists' and 'developers'; sometimes described in the context of 'jobs' versus 'birds'. Against this background the conservation movement has often sought to argue the case for protecting the threatened species, habitats and sites from a standpoint of '**win at all costs**'. The scientific case is marshalled to show how devastating the wildlife losses will be. Arguments may also dispute the claims of the developer not only from an environmental standpoint but also an economic one. Once the battle lines are drawn, the case is heard and a judgement favouring one side or the other is made. Compromise, such as there might be, seldom results in adequate compensation for lost habitat and the usual outcome is a further loss and degradation of what the conservationist considers 'the most precious wildlife areas'.

Whilst these 'battles' may continue to occur, as integrated approaches to management are considered, other more effective scenarios will, it is hoped, develop. In these situations the 'developer' and conservationist will get together and agree the issues, options and possible outcomes against a proposal or series of proposals. There may be recognition on the part of the developer that some options may not be in the best interests of the development. From a nature conservation point of view, on the other hand, new opportunities may evolve which makes protecting the existing interest less vital. Thus protecting habitats and species at all costs might be replaced by a scenario which could be described as '**win some, lose some**'. As sea levels rise in many areas, allowing one wildlife interest to be replaced by another, different one, may be the only viable course of action. With this scenario, conservation may gain more by allowing some areas, previously considered sacrosanct, to provide a buffer for storms, sea level rise and other 'threats' to human use and enjoyment.

REFERENCES

Aartolahti, T., 1973. Morphology, vegetation and development of Rokuanvaara, an esker and dune complex in Finland. *Fennia*, **27**, 1-53.

Aberle, B., 1990. *The Biology, Control and Eradication of Introduced Spartina (Cordgrass) Worldwide and Recommendations for its Control in Washington*. Washington State Department of Natural Resources, Washington.

Adam, P., 1978. Geographical variation in British saltmarsh vegetation. *Journal of Ecology*, **66**, 339-366.

Adam, P., 1981. Vegetation of British saltmarshes. *New Phytologist*, **88**, 143-196.

Adam, P., 1990. *Saltmarsh Ecology*. Cambridge University Press, Cambridge.

Allen, J.R.L., 1992. Tidally influenced marshes in the Severn Estuary, Southwest Britain. In: *Saltmarshes: Morphodynamics, Conservation and Engineering Significance*, eds., J.R.L. Allen & K. Pye, Cambridge University Press, Cambridge, 123-147.

Angus, S. & Elliot, M.M., 1992. Erosion in Scottish machair with particular reference to the Outer Hebrides. In *Coastal Dunes, Geomorphology, Ecology and Management for Conservation*, eds., R.W.B. Carter, T.G.F. Curtis & M.J. Sheehy-Skeffington, A.A Balkema, Rotterdam, 93-112.

Anon, 1977. *Orford Ness, A Selection of Maps Mainly by John Norden*, Presented to J.A. Steers, Heffer & Sons, Cambridge.

Anon, 1992a. *European Workshop on Coastal Zone Management*. Held in Poole, Dorset 21-27 April 1991. Countryside Commission, Cheltenham.

Anon, 1992b. *European Coastal Conservation Conference 1991, proceedings*. Scheveningen/The Hague, Holland 19-21 November 1991. EUCC/Dutch Ministry of Agriculture, Nature Management & Fisheries, The Netherlands, The Hague/Leiden.

Anon, 1992c. *Towards sustainability, Fifth Environmental Action Programme*, European Commission, COM(92)23, 27th March 1992.

Anon, 1996b. *The Pan-European Biological and Landscape Diversity Strategy - a Vision for Europe's Natural Heritage*. Council of Europe, UNEP & European Centre for Nature Conservation.

Anon, 1999. *Goose Populations of the Western Palearctic: a Review of Status and Distribution*. Wetlands International Publication, **48**, National Environmental Research Institute, Denmark and Wetlands International.

Bailey, R.S. 1991. *The Interaction Between Sandeels and Seabirds - a Case History at Shetland*. International Council for the Exploration of the Sea, Copenhagen. (CM 1991/I:41.).

Baker, J.M., 1979. Responses of salt marsh vegetation to oil spills and refinery effluents. In: *Ecological Processes in Coastal Environments*, eds., R.L. Jefferies & A.J. Davy, Blackwell Scientific Publishers, Oxford, 529-542.

Baker, J.M., Adam, P. & Gilfillan, E., 1994. *Biological Impacts of Oil Pollution: Saltmarshes*. International Petroleum Industry Environmental Conservation Association, Report series Vol. **6**.

Bakker, J.P., 1985. The impact of grazing on plant communities, plant populations and soil conditions on saltmarshes. In: *Ecology of Coastal Vegetation*, eds., W.G. Beeftink, J. Rozema & A.H.L. Huiskes. *Vegetatio*, **62**, 391-398.

Bakker, Th.W.M., 1998. Fifty years of recreational planning in the Mijendel dunes near the Hague. In: *Coastal Dunes Recreation and Planning*, ed., J.M. Drees, European Union for Coastal Conservation, Leiden, 11-17.

Bakker, Th.W.M., Jungerius, P.D. & Klijn, J.A., 1990. *Dunes of the European coasts - geomorphology, hydrology and soils*. Catena Supplement **18**, Cremlingen, Holland.

Bakran-Petricioi, T., Petricioli, D., Požar-Domac, A. & Juračić, M., 1996. Importance of preserved natural habitats along the Croatian coast (Adriatic Sea). In: *Partnership in Coastal Zone Management*, eds., J. Taussik & J. Mitchell, Samara Publishing, Cardigan, 195-198.

Balevičienė, J., Lazdauskaitė, Ž. & Rašomavičius, V., 1995. Conservation of botanical diversity in the Lithuanian coastal region. In: *Coastal Conservation and Management in the Baltic Region,* proceedings of the EUCC - WWF conference, 3-7 May 1994, Riga - Klaipėda - Kaliningrad, eds., V. Gudelis, R. Povilanskas & A. Roepstorff, European Union for Coastal Conservation, Leiden, The Netherlands, 43-48

Banks, B., Beebee, T.J.C. & Cooke, A.S., 1994. Conservation of the natterjack toad *Bufo calamita* in Britain over the period 1970-1990 in relation to site protection and other factors. *Biological Conservation*, **67**, 111-118.

Barne, J.H., Robson, C.F., Kaznowska, S.S. & Doody, J.P., 1995/8. *Coasts and Seas of the United Kingdom.* 16 volumes, Joint Nature Conservation Committee, Peterborough.

Barnes, R.S.K., 1974. *Estuarine Biology.* Studies in Biology, No **49**, Edward Arnold, London.

Barnes, R.S.K., 1980. *Coastal Lagoons.* Cambridge University Press, Cambridge.

Barnes, R.S.K., 1996. European coastal lagoons: values and threats. In: *Management of Coastal Lagoons in Albania*, ed., R.M.H. Tekke, European Union for Coastal Conservation, Leiden, 141-146.

Barrère, P., 1992. Dynamics and management of the coastal dunes of Landes, Gascony, France. In: *Coastal Dunes: Geomorphology, Ecology and Management: Proceedings of the Third European Dune Congress, Galway, Ireland, 17-21 June 1992*, eds., R.W.G. Carter, T.G.F. Curtis & M.J. Sheehy-Skeffington, A.A. Balkema, Rotterdam, 25-32.

Bate, G.C. & Dobkins, G.S., 1992. The interactions between sand aeolianite and vegetation in a large transgressive dune sheet. In: *Coastal Dunes: Geomorphology, Ecology and Management: Proceedings of the Third European Dune Congress, Galway, Ireland, 17-21 June 1992*, eds., R.W.G. Carter, T.G.F. Curtis & M.J. Sheehy-Skeffington, A.A. Balkema, Rotterdam, 139-152.

Beatley, T., Brower, D.J. & Schwab, A.K., 1994. *An Introduction to Coastal Zone Management.* Island Press, Washington DC., USA.

Beatty, J., 1992. *Sula: the Seabird-hunters of Lewis.* Michael Joseph, London.

Beeftink, W.G., 1975. The ecological significance of embankment and drainage with respect to the vegetation of the south-west Netherlands. *Journal of Ecology*, **63**, 523-558.

Beeftink, W.G., 1977. Salt-marshes. In: *The Coastline*, ed., R.S.K. Barnes, John Wiley & Sons, Chichester, 93-122.

Bennett, C.M., 1996. Saltmarsh management on an amenity beach of international nature conservation interest at Southport, Merseyside. *Aspects of Applied Biology*, **44**, 301-306.

Bennett, G., 1991. *Towards a European Ecological Network.* Institute of European Environmental Policy, The Netherlands.

Bennett, T., 1984. *Coastal/marine Habitat Change Survey of North Cornwall.* Unpublished Report to the Nature Conservancy Council, Peterborough.

Berry, L.A., 1961. *The Saltmarshes of the Ribble Estuary.* B.Sc. Dissertation, University of Liverpool.

Bignal, E., Bignal, S. & McCraken, D., 1997. The social life of the Chough. *British Wildlife*, **8/6**, 373-383.

Bignal, E.M. & Curtis, D.J. eds., 1989. *Chough and Land-use in Europe.* Scottish Chough Study Group, Argyll.

Binggeli, P., Eakin, M., Macfadyen, A., Power, J. & McConnell, J., 1992. Impact of the alien sea buckthorn (*Hippophae rhamnoides* L.) on sand dunes ecosystems in Ireland. In: *Coastal Dunes: Geomorphology, Ecology and Management: Proceedings of the Third European Dune Congress, Galway, Ireland, 17-21 June 1992*, eds., R.W.G. Carter, T.G.F. Curtis & M.J. Sheehy-Skeffington, A.A. Balkema, Rotterdam, 325-337.

Bird, E.C.F., 1984. *Coasts - an Introduction to Coastal Geomorphology*. 3rd Edition, Basil Blackwell, Oxford.

Bird, E.C.F., 1985. *Coastline Changes: a Global Review*. John Wiley & Sons, Chichester.

Bird, E.C.F., 1993. *Submerging Coasts: the Effects of Rising Sea Levels on Coastal Environments*, John Wiley & Sons, Chichester.

Bird, E.C.F., 1996. *Beach Management*, John Wiley & Sons, Chichester.

Black Sea Environmental Programme, 1997. *Black Sea Geographic Information System*, Version 2.0, April 1997, CD-ROM.

Blondel, J. & Aronson, J., 1999. *Biology and Wildlife of the Mediterranean Region*. Oxford University Press, Oxford.

Boesch, D.F., 1996. Science and management in four U.S. coastal ecosystems dominated by land-ocean interactions. *Journal of Coastal Conservation*, **2**, 103-114.

Boorman, L.A., 1992. The environmental consequences of climate change on British saltmarsh vegetation. *Wetlands Ecological Management*, **2**, 11-21.

Boorman, L.A. & Fuller, R.M., 1977. Studies on the impact of paths on the dune vegetation at Winterton, Norfolk, England. *Biological Conservation*, **12**, 203-216.

Boorman, L.A., Goss-Custard, J.D. & McGrorty, S., 1989. *Climate Change, Rising Sea-Level and the British Coast*. Natural Environment Research Council, HMSO, London.

Boston, K.G., 1981. The introduction of *Spartina townsendii* (s.l.) to Australia. *Occasional Paper - Melbourne State College*, **6**, 1-57.

Boyd, H., 1992. Arctic summer conditions and British knot numbers: an exploratory analysis. In: *The Migration of Knots*, eds., T. Piersma & N. Davidson, Wader Study Group Bulletin, **64**, 144-154.

Boyd, H. & Pirot, J.-Y., eds., 1989. *Flyways and Reserve Networks for Waterbirds*. International Waterfowl and Wetlands Research Bureau, Special Publication, **9**, Slimbridge.

Brandt, E. & Christensen, S.N., 1994. *Danske Klitter en Oversigtlig Kortlaegning* (2 volumes), Miljøministeriet Skov-og Naturstyrelsen.

Britton, R.H. & Johnson, A.R., 1987. An ecological account of a Mediterranean salina: the Salin-De-Giraud, Camargue (S. France). *Biological Conservation*, **42**: 185-230.

Brothers, N.P., 1991. Albatross mortality and associated bait loss in the Japanese long-line fishery in the southern Ocean. *Biological Conservation*, **55**, 255-268.

Brown, A.C. & McLachlan, A., 1990. *Ecology of Sandy Shores*. Elsevier, Amsterdam.

Bull, K.R., Every, W.J., Freestone, P., Hall, J.R. & Osborn, D., 1983. Alkyl lead pollution and bird mortality on the Mersey Estuary 1979-1981. *Environmental Pollution*, **31**, (Series A), 239-259.

Burd, F., 1989. *The Saltmarsh Survey of Great Britain. An Inventory of British Saltmarshes*. Research & survey in nature conservation, **17**, Nature Conservancy Council, Peterborough.

Burd, F., 1992. *Erosion and Vegetation Change on the Saltmarshes of Essex and North Kent between 1973 and 1988*. Research & survey in nature conservation, **42**, Nature Conservancy Council, Peterborough.

Burd, F., 1995. *Managed Retreat: a Practical Guide*. English Nature, Campaign for a living coast, Peterborough.

Burden, R.F. & Randerson, P.F., 1972. Quantitative studies of the effects of human trampling on vegetation as an aid to the management of semi-natural areas. *Journal of Applied Ecology*, **9**, 339-357.

Cadbury, J., & Ausden, M., in Press. Bird communities of coastal shingle and lagoons. In: *Ecology & Geomorphology of Coastal Shingle*, eds., J.R. Packham, R.E. Randall, R.S.K. Barnes & A. Neal, Westbury Academic & Scientific Publishing, .

Cadwallader, D.A. & Morley, J.V., 1971. Further experiments on the management of saltings pasture for Wigeon (*Anas penelope* L.) conservation at Bridgwater Bay National Nature Reserve, Somerset, *Journal of Applied Ecology*, **10**, 161-166.

Cadwalladr, D.A., Owen, M., Morley, J.V. & Cook, R.S., 1972. Wigeon (*Anas penelope*) conservation and salting pasture management at Bridgwater Bay National Nature Reserve, Somerset, *Journal of Applied Ecology*, **9**, 117-126.

Carey, A.E. & Oliver, F.W., 1918. *Tidal Lands, a Study in Shore Problems*. Blackie and Son, London.

Carr, A.P., 1970. The evolution of Orfordness, Suffolk, before 1700 AD: geomorphological evidence. *Zeitschrift für Geomorphologie, New Series*, **14**, 289-300.

Carter, R.W.G., 1988. *Coastal Environments: an Introduction to the Physical Ecological and Cultural Systems of Coastlines*. Academic Press, London.

Carter, R.W.G. & Wilson, P., 1990. The geomorphology, ecological and pedological development of coastal foredunes at Magilligan Point, Northern Ireland. In: *Coastal Dunes - Form and Process*, eds., K. Nordstrom, N. Psuty & B. Carter, John Wiley & Sons, Chichester, 129-157.

Carter, R.W.G. & Woodroffe, C.D., eds., 1994. *Coastal Evolution, Late Quaternary Shoreline Morphodynamics*, Cambridge University Press.

Cencini, C., Marchi, M., Torresani, S. & Varani, L., 1988. The impact of tourism on Italian deltaic coastlands: four case studies. *Ocean & Shoreline Management*, **11**, 353-374.

Chapman, V.J., 1938. Studies in saltmarsh ecology, I - III. *Journal of Ecology*, **26**, 144-179.

Chapman, V.J., 1939. Studies in saltmarsh ecology IV. *Journal of Ecology*, **27**, 181-201.

Chapman, V.J., 1941. Studies in saltmarsh ecology, VIII. *Journal of Ecology*, **29**, 69-82.

Chapman. V.J., 1959. Studies in saltmarsh ecology. IX. Changes in saltmarsh vegetation at Scolt Head Island. *Journal of Ecology*, **67**, 619-639.

Chapman, V.J., 1964. *Coastal Vegetation*. Pergamon Press, Oxford.

Chapman, V.J., 1974. *Saltmarshes and Salt Deserts of the World*. Leonard Hill, London.

Charman, K., 1990. The current status of *Spartina anglica* in Britain. In: *Spartina anglica - a Research Review*, eds., A.J. Gray & P.E.M. Benham, Institute of Terrestrial Ecology, HMSO, London, 11-14.

Christiansen, C., Dalsgaard, K., Møller, J.T. & Bowman, D., 1990. Coastal dunes in Denmark. Chronology in relation to sea level. *Catena Supplement*, **18**, 61-70.

Chung, Chung-Hsin, 1990. Twenty-five years of introduced *Spartina anglica* in China. In: *Spartina anglica - a Research Review*, eds., A.J. Gray & P.E.M. Benham, Institute of Terrestrial Ecology, HMSO, London, 72-76.

Clark, R.B., 1984. Impact of Oil Pollution on Seabirds. *Environmental Pollution*, A **33**, 1-22.

Coccossis, 1997. *Integrated Coastal Area and River Basin Management*. ICARM Technical report Series No **1**, UNEP, Priority Actions Programme.

Coles, S.M., 1979. Benthic microalgal populations on intertidal sediments and their role as precursors to saltmarsh development. In: *Ecological Processes in Coastal Environments*, eds., R.L. Jefferies & A.J. Davy, Blackwell Scientific Publications, Oxford, 25-52.

Cooke, J.A. & Gray, S., 1984. Nature conservation on the Durham coast - the future. *The Vasculum*, Vol. **69, 3**, 84-86.

Cooper, J.A.G., 1994. Lagoons and microtidal coasts. In:. *Coastal Evolution, Late Quaternary Shoreline Morphodynamics*, eds., R.W.G. Carter, & C.D. Woodroffe, Cambridge University Press, Cambridge, 219-265.

Cori, B., 1999. Spatial dynamics of Mediterranean coastal regions. *Journal of Coastal Conservation*, **5**, 105-112.

Corre, T-J., 1993. Dry coastal ecosystems of Southern France. In: *Dry Coastal Ecosystems, 2A Polar Regions and Europe*, ed., E. van der Maarel, Elsevier, Amsterdam, 309-375.

Correia, F., Dias, J.A., Boski, T. & Ferreira, O., 1996. The retreat of the eastern Quarteira cliffed coast (Portugal) and its possible causes. In: *Studies in European Coastal Management*, eds., P.S. Jones, M.G. Healy & A.T. Williams, Samara Publishing, Cardigan, 129-136.

Costanza, R., W.M. Kemp & W.R. Boynton, 1993. Predictability, scale and biodiversity in coastal and estuarine ecosystems: implications for management. *Ambio*, **22**, 88-96.

Council of Europe, 1998. *Coastal Zones - Towards Sustainable Management*. Naturopa, **88**, Centre Naturopa, Strasbourg.

Daiber, F.C., 1982. *Animals of the Tidal Marsh*. Van Nostrand Reinhold, New York.

Daiber, F.C., 1986. *Conservation of Tidal Marshes*. Van Nostrand Reinhold, New York.

Dalby, R., 1957. Problems of land reclamation: 5, Saltmarsh in the Wash. *Agricultural Review, London*, **2**, 31-37.

Dardeau, M.R., Modlin, R.F., Schroeder, W.W. & Stout, J.P., 1992. Estuaries. In: *Biodiversity of the Southeastern United Sates: Aquatic Communities*, eds., C.T. Hackney, S.M. Adams & W.H. Martin, John Wiley, New York, 615-755.

Dargie, T.C.D., 1993. *Sand Dune Vegetation Survey of Great Britain. Part 2 - Scotland*. Peterborough, Joint Nature Conservation Committee.

Dargie, T.C.D., 1995. *Sand Dune Vegetation Survey of Wales. Part 3 - Wales*. Peterborough, Joint Nature Conservation Committee.

Davidson, I., 1999. Migratory bird conservation in the Americas. *Wetlands, the Newsletter of Wetlands International*, No. **8**, 10-11.

Davidson, N.C. 1991. Breeding waders on British estuarine grasslands. *Wader Study Group Bulletin*, **61**, Supplement: 36-41.

Davidson, N.C., Laffoley, D. d'A. & Doody, J.P., 1995. Land claim in British estuaries: changing patterns and conservation implications. In: *The Changing Coastline*, ed., N.V. Jones, Coastal Zone Topics: process, ecology & management, **1**, Joint Nature Conservation Committee, Peterborough, 68-80.

Davidson, N.C., Laffoley, D. d'A., Doody, J.P., Way, L.S., Gordon, J., Key, R., Drake, C.M., Pienkowski, M.W., Mitchell, R. & Duff, K.L. 1991. *Nature Conservation and Estuaries in Great Britain*. Nature Conservancy Council, Peterborough.

Davidson, N.C. & Stroud, D.A., 1996. Conserving international coastal habitats networks on migratory waterfowl flyways, *Journal of Coastal Conservation* **2**, 41-54.

Davies, M., 1987. Twite and other wintering passerines on the Wash saltmarshes. In: *The Wash and its Environment*, eds., J.P. Doody & B. Barnet, Research & survey in nature conservation, No. **7**, Nature Conservancy Council, Peterborough, 123-132.

Davis. P. & Moss, D., 1984. *Spartina* and waders the Dyfi estuary. In: *Spartina anglica in Great Britain*, ed., J.P. Doody, Focus on nature conservation, No. **5**, Nature Conservancy Council, Attingham Park, 37-40.

Day, J.W. Jr., Martin, J.F., Cardoch, L. & Templet, P.H., 1997. System functioning as a basis for sustainable management of deltaic ecosystems. *Journal of Coastal Management*, **25**, 115-153.

Day, J.W. Jr., Rismondo, A., Scarton, F., Arc, D. & Cecconi, G., 1998. Relative sea level rise and Venice lagoon wetlands. *Journal of Coastal Conservation*, **4**, 27-34.

de Bonte, A.J., Boosten, A., van der Hagen, H.G.J.M. & Sýkora, K.V., 1999. Vegetation development influenced by grazing in the coastal dunes near The Hague, The Netherlands. *Journal of Coastal Conservation*, **5**, 59-68.

de Raeve, F., 1989. Sand dune vegetation and management dynamics. In: *Perspectives in Coastal Dune Management*, eds., F. van der Meulen, P.D. Jungerius & J. Visser, SPB Academic Publishing, The Hague, 99-109.

de Ruig, J.H.M., 1998. Coastline management in the Netherlands: human use versus natural dynamics. *Journal of Coastal Conservation*, **4**, 127-134.

de Ruig, J.H.M. & Hillen, R., 1997. Developments in Dutch coastline management: conclusions from the second governmental coastal report. *Journal of Coastal Conservation*, **3**, 203-210.

Delbaere, B.C.W., ed., 1999. Facts and Figures on Europe's Biodiversity: State and Trends 1998-1999. European Centre for Nature Conservation, Tilburg, The Netherlands.

Dijkema, K..S., 1990. Salt and brackish marshes around the Baltic Sea and adjacent parts of the North Sea: their vegetation and management. *Biological Conservation*, **51**, 191-209.

Dijkema, K.S., ed., 1984. *Salt marshes in Europe.* Nature and environment series, No. **30**, Council of Europe, Strasbourg.

Dijkema, K..S. & Wolff, W.J., 1983. *Flora and Vegetation of the Wadden Sea Islands and Coastal Areas.* Report **9** of the Wadden Sea Working Group, Stitching Veth tot Steun aan Waddenonderzoek, Leiden, 305-308.

Doarks, C. 1990. *Changes in the Flora of Grazing Marsh dykes in Broadland Between 1972-74 and 1977-79.* Nature Conservancy Council (England Field Unit Report), Peterborough.

Dobbs, C.G., 1939. The vegetation of Cape Napier, Spitzbergen. *Journal of Ecology*, **27**, 126-148.

Doody, J.P., 1981. The conservation of the semi-natural vegetation of the Magnesian Limestone; 3, The Durham coast. *The Vasculum*, **66**, 17-28.

Doody. J.P. ed., 1984. *Spartina anglica in Great Britain.* Focus on nature conservation, No. **5**, Nature Conservancy Council, Attingham Park.

Doody, J.P., 1987. The impact of "reclamation" on the natural environment of the Wash. In: *The Wash and its environment*, eds. P. Doody & B. Barnett, Research &survey in nature conservation, No **7**, Nature Conservancy Council, Peterborough, 165-172

Doody, J.P., 1989. Management for nature conservation. *Proceedings Royal Society of Edinburgh*, **96B**. 247-266.

Doody., J.P. ed., 1991. *Sand Dune Inventory of Europe*, Joint Nature Conservation Committee, Peterborough, the European Union for Coastal Conservation, Leiden.

Doody, J.P., 1992. Sea defence and nature conservation: threat or opportunity. *Journal of Aquatic Conservation*, **2**, 275-283.

Doody, J.P., 1995. Infrastructure development and other human influences on the coastline of Europe. In: *Coastal Management and Habitat Conservation*, eds., A.H.P.M. Salman, H. Berends & M. Bonazountas. Proceedings of the 4th EUCC Congress, Vol. **I**, Marathon, Greece, April, 1993, European Union for Coastal Conservation, Leiden, 133-182.

Doody, J.P., 1996. Management and use of dynamic estuarine shorelines. In: *Estuarine Shores, Evolution, Environments and Human Alterations*, eds., K.F. Nordstrom, & C.T. Roman, Wiley, Chichester, 421-434.

Doody, J.P., 1999. Saltmarshes and sand dunes - natural or not. In: *Ecology and Landscape Development - A History of the Mersey Basin*, ed., E.F. Greenwood, Liverpool University Press, National Museums & Galleries on Merseyside, 167-174.

Doody, J.P., Johnson, C. & Smith, B., eds., 1993. *Directory of the North Sea Coastal Margin*, Joint Nature Conservation Committee, Peterborough.

Doody, J.P., Pamplin, C., Gilbert, C. & Bridge, L., 1999. *European Union Demonstration programme on Integrated Coastal Zone Management - Theme F, INFORMATION.* Report to the European Environment Agency, Copenhagen.

Doody, P. & Salman, A., 1998. The European Coastal and Marine Ecological Network. In: *The Green Backbone of Central and Eastern Europe*, ed., P. Nowicki, series on Man and Nature, **3**, European Centre for Nature Conservation, Tilburg, the Netherlands, 130-147.

Dorset Coast Forum, 1999. *Dorset Coast Strategy, Proposals for Policy and Action for Dorset's Coast.* Dorset Coast Forum.

Driscoll, R.J., 1983. Broadland dykes: the loss of an important wildlife habitat. *Transactions of the Norfolk and Norwich Naturalists' Society*, **26** (3), 120-172.

Duffey, E., 1967. An assessment of dune invertebrate faunas in habitats vulnerable to public disturbance. In: *The biotic Effects of Public Pressures on the Environment*, ed., E. Duffey, Symposium **No. 3**, Monks Wood Experimental Station, Cambridge.

Duffey, E., 1968. An ecological analysis of the spider fauna of sand dunes. *Journal of Animal Ecology*, **37**, 641-674.

Duffey, E., 1970. Habitat selection by spiders on a saltmarsh in Gower. *Nature in Wales*, **12**, 15-23.

Dugan, P., ed., 1993. *World Conservation Atlas: Wetlands in Danger*. Mitchell Beazley, London.

Dyer, K.R., 1972. *Estuaries: a Physical Introduction*. Wiley-Interscience, London.

East Durham Task Force, 1995. *Turning the Tide: Restoring the Coast of County Durham. A Project for the Millennium*. East Durham Task Force, Durham County Council.

Eisma, D., 1978. Stream deposition and erosion by the eastern shore of the Aegean. In: *The Environmental History of the Near and Middle East Since the Last Ice Age*, ed., W.C. Brice, Academic Press, London, 67-81.

Ellis, S., 1997. The Northern brown Argus in North-east England. *British Wildlife*, Vol. **9**, **1**, 22-27.

Environment Agency, 1999. *A Guide to the Understanding and Management of Saltmarshes*. R&D Note **324**. Chapter 10, updated 1999, Environment Agency, Bridgwater, Somerset, 307-326.

European Commission, 1996. *Demonstration Programme on Integrated Management of Coastal Zones*. Information document, European Commission Services, XI/79/96.

European Commission, 1999a. *Towards a European Integrated Coastal Zone Management (ICZM) Strategy - General Principles and Policy Options. A Reflection Paper*. Directorates-General Environment, Nuclear Safety and Civil Protection, Fisheries, Regional Policies and Cohesion.

European Commission, 1999b. *Lessons from the European Commission's Demonstration Programme on Integrated Coastal Zone Management (ICZM)*. Directorates-General Environment, Nuclear Safety and Civil Protection, Fisheries, Regional Policies and Cohesion.

European Commission, 1999c. *Interpretation Manual of European Habitats. Natura 2000*. European Commission, DG Environment, Brussels.

Evans, G. & Collins, M., 1987. Sediment supply and deposition in the Wash. In: *The Wash and its Environment,* eds., P. Doody & B. Barnett. Research and Survey in Nature Conservation, No. **7**, Nature Conservancy Council, Peterborough, 48-63.

Evans, P., 1990. *Whales*. Whittet Books, London.

Evans, P.R., 1984. Use of the herbicide "Dalapon" for control of *Spartina* encroaching on intertidal mudflats: beneficial effects on shorebirds. *Colonial Waterbirds*, **9**, 171-175.

Evans, P.R., Herdson, D.M., Knights, P.J. & Pienkowski, M.W., 1979. Short-term effects of reclamation of part of Seal Sands, Teesmouth, on wintering waders and shelduck. *Oecologia*, **41**, 183-206.

Fabbri, P., 1997. Tourism, recreation and coastal dunes of the Mediterranean. In: *Coastal Dunes, Recreation and Planning*, ed., J.M. Drees, EUCC, Leiden, The Netherlands, 18-20.

Fairbridge, R.W., 1980. The estuary: its definition and geodynamic cycle. In: *Chemistry and Biogeochemistry of Estuaries*, eds., E. Olausson & I. Cato, New York, Wiley, 1-35.

Farrell, L. & Randall, R.E., 1992. The distribution of ***Mertensia maritima***, (L.) (Grey oyster plant) in Ireland. *Irish Naturalist Journal*, **24**, 135-140.

Favennec, J., 1998. The dunes of the Atlantic coast of France, typology and management. In: *Coastal Dunes, Management, Protection and Research*, ed., C.H. Ovesen, report from a European seminar, Skagen, Denmark, National Forest and Nature Agency, Geological Survey of Denmark and Greenland, 130-138.

Ferry, B.W., in Press. Vegetation of Dungeness shingle: aspects of biodiversity. In: *Ecology & Geomorphology of Coastal Shingle*, eds., J.R. Packham, R.E. Randall, R.S.K. Barnes & A. Neal. Westbury Academic & Scientific Publishing, Otley, West Yorkshire.

Ferry, B. & Waters, S., 1985. *Dungeness: Ecology and Conservation*. Focus on nature conservation, No. **12**. Nature Conservancy Council, Peterborough.

Ferry, B.W., Waters, S.J.P. & Jury, S.L., 1989. *Dungeness: the Ecology of a Shingle Beach*. Published as special issue of the *Botanical Journal of the Linnean Society*, **101**, by Academic Press, London.

Ferry, B.W., Barlow, S.L. & Waters, S.J.P., 1989. The shingle ridge succession at Dungeness. In: *Dungeness - the Ecology of a Shingle Beach*, eds., B.W. Ferry, S.J.P. Waters & S.L. Jury. *Botanical Journal of the Linnean Society*, **101**, 19-30.

Ferry, B., Lodge, N. & Waters, S., 1990. *Dungeness: a Vegetation Survey of a Shingle Beach*. Research & survey in nature conservation, No. **26**. Nature Conservancy Council, Peterborough.

Findon, R., 1985. Human pressures. In: *Dungeness: Ecology and Conservation*, eds., B. Ferry & S. Waters, Focus on nature conservation, **12**, Nature Conservancy Council, Peterborough, 13-24.

Findon, R., 1989, Recent developments at Dungeness. . In: *Dungeness: the Ecology of a Shingle Beach*, eds., B.W. Ferry, S.J.P. Waters & S.L. Jury, *Botanical Journal of the Linnean Society*, **121**, 125-135

Fjelland, E., 1983. *Botanical Conservation Values on Seashores in Tromso. Northern Norway*. Institute of Biology and Geology, University of Tromso.

Fuller, R.M., 1985. An assessment of damage to the shingle beaches and their vegetation. In: *Dungeness Ecology and Conservation*, eds., B. Ferry & S. Waters. Focus on Nature Conservation, **12**, Nature Conservancy Council, Peterborough, 25-42.

Fuller, R.M. & Randall, R.E., 1988 The Orford Shingles, Suffolk, UK - Classic Conflicts in Coastline Management. *Biological Conservation* **47**, 95-114.

Fuller, R.J., Reed, T.M., Buxton, N.E., Webb, A., Williams, T.D. & Pienkowski, M.W., 1986. Populations of breeding waders *Charadriidae* and their habitats on the crofting lands of the Outer Hebrides, Scotland. *Biological Conservation*, **37**, 333-361

Garcia Novo, F., 1997. The ecosystems of Doñana National Park (southwest Spain). In. *The Ecology and Conservation of European Dunes*, eds., F. Garcia Novo, R.M.M. Crawford & Diaz Barradus, C.D., Universidad de Sevilla, Savilla, 149-154.

Garniel, A. & Mierwald, U., 1996. Changes in the morphology and vegetation along the human altered shoreline of the Lower Elbe. In: *Estuarine Shores*, eds., K.F. Nordstrom & C.T. Roman, Wiley, Chichester, 375-396.

Géhu, J.M., 1960. Une site célèbre de la côte Nord Bretonne: Le sillon de Talbert (C-du-N). Observation Phytosociologiques et écologiques, *Bulletin Laboratoire Maritime, Dinnard*, **46**, 93-115.

Géhu, J.M., 1984. 'Saltmarsh loss' Southern Europe. In: *Salt marshes in Europe*. Nature and environment series, ed., K.S. Dijkema No. **30**, Council of Europe, Strasbourg.

Géhu J.M. & Géhu-Franck J., 1993. Dry coastal ecosystems of Belgium and the Atlantic coasts of France. In: *Ecosystems of the World 2A. Dry Coastal Ecosystems - Polar Regions and Europe*, ed., E. van der Maarel, Elsevier, The Netherlands, 307-327.

Giddings, J.L,. 1977. *Ancient Men of the Arctic*. Alfred A. Knoff.

Gimingham, C.H., 1992. *The Lowland Heathland Management Handbook. English Nature Science*, No. **8**, English Nature, Peterborough, UK.

Gjiknuri, L. & Hoda, P., 1996. The ecological evaluation of the most important lagoons in Albania. In: *Management of coastal lagoons in Albania*, ed., R.M.H. Tekke, EUCC, Leiden, 7-20

Goeldner, L., 1999. The German Wadden Sea coast: reclamation and environmental protection. *Journal of Coastal Conservation*, **5**, 23-30.

Goldsmith, F.B., 1977. Rocky cliffs. In: *The Coastline*, ed., R.S.K. Barnes, John Wiley & Sons, Chichester, 237-251.

Gornitz, V., 1993. Mean sea levels changes in the recent past. In: *Climate and Sea Level Change, Observations, Projections and Implications*, eds., R.A. Warrick, E.M. Barrow & T.M.L. Wigley, Cambridge University Press, 25-55.

Goss-Custard & Moser, M., 1990. Changes in the numbers of dunlin (*Calidris alpina*) in British estuaries in relation to changes in the abundance of *Spartina*. In: *Spartina anglica - a Research Review*, eds., A.J. Gray & P.E.M. Benham. Institute of Terrestrial Ecology, HMSO, London, 69-71.

Goutner, V., 1992. Management of Mediterranean Lagoons and Saltmarshes. *Abstracts of papers to the Medmaravis Conference*, Chios, Greece, September 1992.

Granja, H.G. & Soares de Carvalho, G., 1991. The impact of "protection" on structures on the Ofir-Apúlia coastal zone (NW Portugal). *Quaternary International*, **9**, 81-85.

Gray, A.J., 1972. The ecology of Morecambe Bay. V. The Saltmarshes of Morecambe Bay. *Journal of Applied Ecology*, **9**, 207-220.

Gray, A.J., 1977. Reclaimed land. In: *The Coastline*, ed., R.S.K. Barnes, John Wiley & Sons, Chichester, 253-270.

Gray, A.J., 1990. *Spartina anglica* - the evolutionary and ecological background. In: *Spartina anglica - a Research Review*, eds., A.J. Gray & P.E.M. Benham. Institute of Terrestrial Ecology, HMSO, London, 5-10.

Gray, A.J., ed., 1992. *The Ecological Impact of Estuarine Barrages*. British Ecological Society, Ecological Issues No. **3**, Field Studies Council, Shrewsbury.

Gray, A.J. & Scott, R., 1987. Saltmarshes. In: *Morecambe Bay - an Assessment of Present Ecological Knowledge*, eds., N.A. Robinson & A.W. Pringle, Morecambe Bay Study Group, in conjunction with Centre for North West Regional Studies, University of Lancaster, 97-117.

Gray, A.J. & Benham, P.E.M., 1990. *Spartina anglica - a Research Review*. Institute of Terrestrial Ecology, HMSO, London.

Gray, A.J., Raybould, A.F. & Hornby, O., 1999. *A Survey of Non-Spartina anglica species in the Solent*. Internal Report to English Nature, Peterborough.

Green, C., & Penning-Rowsell, E., 1999. Inherent conflicts at the coast. *Journal of Coastal Conservation*, **5**, 153-162.

Green, M., 1997. Review of the state of the World's protected areas. WCPA Symposium: *Protected Areas in the 21st Century: from Islands to Networks*. November 1997, Albany, Western Australia.

Grenon M. & Batisse M., 1989. *The Blue Plan - Futures for the Mediterranean Basin*. UNEP, Oxford University Press.

Griggs, G.B. & Trenhaile A.S., 1994. Coastal cliffs and platforms. In: *Coastal Evolution - Late Quaternary Shoreline Morphodynamics*, eds., R.W.G. Carter & C.D. Woodroffe, Cambridge University Press, 425-450.

Haeseler, V., 1985. Nord-und Ostfriesische Inseln als "Reservate" thermophiler Insekten am Beispiel der Hymenopters Aculeata, *Mitt. dtsch. Ges. allg. angew. Ent.*, **4**, 447-452.

Haeseler, V., 1989. The situation of the invertebrate fauna of coastal dunes and sandy coasts in the western Mediterranean (France, Spain). In: *Perspectives in Coastal Dune Management*, eds., F. van der Meulen, P.D. Jungerius & J. Visser, SPB Academic Publishing, The Hague, 128-131.

Haeseler, V., 1992. Coastal dunes of the southern North Sea as habitats of digger wasps. In: *Coastal dunes: geomorphology, ecology and management: proceedings of the third European Dune Congress, Galway, Ireland, 17-21 June 1992*, eds., R.W.G. Carter, T.G.F. Curtis & M.J. Sheehy-Skeffington, A.A. Balkema, Rotterdam, 381-389.

Hafner, H. & Fasola, M., 1992. The relationship between feeding habitat and colonially nesting Ardeidae. In: *Managing Mediterranean Wetlands and Their Birds*. Proceedings of an IWRB International Symposium, Grado, Italy, eds., M. Finlayson, T. Hollis & T. Davis. IWRB Special Publication, **20**, 194-201.

Handelmann, D., 1998. Dune insects - grasshoppers (Saltatoria) and ants (Formicidae). In: *Coastal Dunes, Management, Protection and Research*, ed., C.H., Ovesen, report from a European seminar, Skagen, Denmark, National Forest and Nature Agency, Geological Survey of Denmark and Greenland, 87-90.

Hansom, J.D., 1999. The coastal geomorphology of Scotland: understanding sediment budgets for effective coastal management. In: *Scotland's living coastline*, eds., J.M. Baxter, K. Duncan, S.M. Atkins & G. Lees, Scottish Natural Heritage, The Stationery Office, London, 34-44.

Harold, R., 1995. Creating wetlands at Holkham. *Enact*, **3**, English Nature, 12-15.

Harris, M.P., 1984. *The Puffin*. T&AD Poyser, Calton, London.

Harrison, D.J., 1996. Resource management of marine and sand and gravel: A European perspective. In: *Partnership in Coastal Management*, eds., J. Taussik & J. Mitchell, Samara Publishing, Cardigan, 15-20.

Harrison, J.M., 1953. *The Birds of Kent*. 2 Volumes, Witherby, London.

Hawkins, S.J.., Allen, J.R., Fielding, N.J., Wilkinson, S.B. & Wallace, I.D., 1999. Liverpool Bay and the estuaries: human impact, recent recovery and restoration. In: *Ecology and Landscape Development: A History of the Mersey Basin*, ed., E.F. Greenwood, Liverpool University Press, National Museums & Galleries on Merseyside, 155-165.

Haynes. F.N., 1984. *Spartina* in Langstone Harbour, Hampshire. In*: Spartina anglica in Great Britain*, ed., J.P. Doody, Focus on nature conservation, No. **5**, Nature Conservancy Council, Attingham Park, 5-10.

Hazelden, J., Loveland, P.J. & Sturdy, R.G., 1986. *Saline Soils in North Kent*. Soil Survey Special Survey, **14**, Harpenden.

Hearn, K.A., 1995. Stock grazing of semi-natural habitats on National Trust land. In: *The National Trust and Nature Conservation: 100 Years On*, eds., D.J. Bullock & H.J. Harvey. *Biological Journal of the Linnean Society*, **56** (Suppl.), 25-37.

Heikkinen, O. & Tikkanen, M., 1987. The Kalajoki dune field on the west coast of Finland, *Fennia*, **165**: 2, 241-267.

Hill, D. & Makepeace, P., 1989. Population trends in bird species at Dungeness, Kent. In: *Dungeness: the Ecology of a Shingle Beach*, eds., B.W. Ferry, S.J.P. Waters & S.L. Jury, *Botanical Journal of the Linnean Society*, **121**, 137-151.

Hill, M. I., 1988. *Saltmarsh Vegetation of the Wash - an Assessment of Change from 1971 to 1987*. Research & survey in nature conservation, **13**, Nature Conservancy Council, Peterborough.

Hinrichsen, D., 1990. *Our Common Seas: Coasts in Crisis*. Earthscan, London.

Hinrichsen, D., 1998. *Coastal Waters of the World, Trends, Threats, and Strategies*. Island Press, Washington DC.

Hopkins, J.J., 1979. The alien *Carpobrotus edulis* - a threat to the Lizard flora. *Proceedings of the Lizard Field Studies Club*, **6**, 14-15.

Hopkins, J.J., 1983. *Studies of the Historical Ecology, Vegetation and Flora of the Lizard District, Cornwall*. 2 vols. PhD Thesis, University of Bristol.

Houston, J., 1992. Blowing in the wind. *Landscape Design*, 25-29.

Hubbard, J.C.E. & Stebbings. R.E., 1967. Distribution, dates of origin and acreage of *Spartina townsendii* salt marshes in Great Britain. *Transactions of the Botanical Society of the British Isles*, **7**, 1-7.

Huisman, L. & Oltshoorn, T.N., 1983. *Artificial Groundwater Recharge*, Pitman, Boston.

Humphries, L., 1996. A coastal morphology classification system for beaches of the Co. Durham coast modified by the addition of colliery spoil. In: *Partnership in Coastal Zone Management*, eds., J. Taussik & J. Mitchell, Samara Publishing Limited, Cardigan, 317-325.

Hundt, R., 1993. Dry coastal ecosystems of the Baltic coast of Germany In: *Ecosystems of the World 2A. Dry Coastal Ecosystems - Polar Regions and Europe*, ed., E. van der Maarel, Elsevier, The Netherlands, 165-181.

Ibàñez, C., Caicio, A., Day, J.W. & Curcó, A., 1997. Morphological development, relative sea level rise and sustainable management of water and sediment in the Ebre Delta, Spain, *Journal of Coastal Conservation*, **3**, 191-202.

Innocenti, L. & Pranzini, E., 1993. Geomorphological evolution and sedimentology of the Ombrone river delta, Italy. *Journal of Coastal Research*, **9**/2, 481-493.

International Union for the Conservation of Nature, 1992. *Protected Areas of the World: a Review of National Systems*; Vol. **1**, Indomalaya, Oceania, Australia and Antarctica; Vol. **2**, Palaearctic and Vol. **3**, Afrotropical. International Union for the Conservation of Nature, Gland, Switzerland, and Cambridge, UK.

International Union for the Conservation of Nature, 1998. *Coastal Conservation Policy. Guidelines for the Political Actors*. IUCN, Commission on Environmental, Economic and Social Policy, Coastal Working Group.

Isermann, M. & Cordes, H., 1992. Changes in dune vegetation on Spiekeroog (East Friesian islands) over a 30 year period. In: *Coastal dunes: geomorphology, ecology and management: proceedings of the third European Dune Congress, Galway, Ireland, 17-21 June 1992*, eds., R.W.G. Carter, T.G.F. Curtis & M.J. Sheehy-Skeffington, A.A. Balkema, Rotterdam, 381-389.

Ishizuka, K., 1975. Maritime vegetation. In: *The Flora and Vegetation of Japan*, ed., M. Numata, Elsevier, Amsterdam, 151-172

Jeffrey, D.W., 1977. *North Bull Island, Dublin Bay - a Modern Coastal Natural History*. The Royal Dublin Society, Dublin.

Jelgersma, S., de Jong, J., Zagwijn, W.H. & Regteren Altena, J.F. van, 1970. The coastal dunes of the western Netherlands, geology, vegetational history and archaeology. *Med. Rijks Geol. Dienst N.S.* **21**, 93-167.

Jensen, A., 1993. Dry coastal ecosystems of Denmark. In: *Ecosystems of the World 2A. Dry coastal ecosystems - Polar regions and Europe*, ed., E. van der Maarel, Elsevier, The Netherlands, 183-196.

Jensen, F., 1995. The Danish experience in recreation and planning in and around coastal dunes. In: *Coastal Dunes Recreation and Planning*, ed., J.M. Drees, European Union for Coastal Conservation, Leiden, 35-42.

Johnson, A., 1992. The west Mediterranean population of greater flamingo: is it at risk? In: *Managing Mediterranean Wetlands and Their Birds*, eds., M. Finlayson, T. Hollis & T. Davis. Proceedings of an IWRB International Symposium, Grado, Italy, IWRB Special Publication, **20**, 215-219.

Johnson, A.R., 1982. Construction of a breeding island for flamingos in the Camargue. In: *Managing Wetlands and Their Birds*, ed., D.A. Scott, IWRB, Slimbridge.

Jones, P.S. & Etherington, J.R., 1992. Autecological studies on the rare orchid *Liparis loeselii* and their application to the management of dune slack ecosystems in South Wales. In: *Coastal Dunes: Geomorphology, Ecology and Management: Proceedings of the Third European Dune Congress, Galway, Ireland, 17-21 June 1992*, eds., R.W.G. Carter, T.G.F. Curtis & M.J. Sheehy-Skeffington, A.A. Balkema, Rotterdam, 299-312.

Jungerius, P.D., Koehler, H., Kooijman, A.M. Mücher, H.J. & Graefe, U., 1995. Response of vegetation and soil ecosystems to mowing and sod removal in the coastal dunes 'Zwanenwater' The Netherlands. *Journal of Coastal Conservation*, **1**, 3-16.

Kamps L.F., 1962. *Mud distribution and land reclamation in the eastern Wadden shallows. Rijkswaterstaat Communications* No. **5**, Den Hague.

Kelley, D., 1986. Bass nurseries on the west coast of the UK. *Journal of the Marine Biological Association*, **66**, 439-464.

Kestner, F.T.J., 1975. The loose-boundary regime of the Wash. *Geographical Journal*, **151**, 388-515.

Kirby, P., 1992. *Habitat Management for Invertebrates: a Practical Handbook*. Joint Nature Conservation Committee, Peterborough, RSPB, Sandy, Bedfordshire.

Kirby, J., West, R., Scott, D., Davidson, N., Piersma, T., Hötker, H. & Stroud, D.A., 1999 (Draft). *Atlas of Wader Populations in Africa and Western Eurasia*. Wader Study Group, Netherlands.

Klijn, J.A., 1990. The younger dunes in the Netherlands: chronology and causation. . In: *Dunes of the European Coasts, Geomorphology-Hydrology-Soils*, eds., Th. W. Bakker, P.D. Jungerius & J.A. Klijn, *Catena Supplement*, **18**, 89-100.

Lammerts, E.J., Grootjans, A., Stuyfzand, P & Sival, F., 1995. Endangered dune slack plants: gastronomers in need of mineral water. In: *Coastal Management and Habitat Conservation*, eds., A.H.P.M. Salman, H. Berends & M. Bonazountas. Proceedings of the 4th EUCC Congress, Vol. I, Marathon, Greece, April, 1993, European Union for Coastal Conservation, Leiden, The Netherlands, 383-369.

Lampe, R., 1996. Shoreline changes in the Bodden coast of northeastern Germany. In: *Estuarine Shores*, eds., K.F. Nordstrom & C.T. Roman, Wiley, Chichester, 63-88.

Lane, P.A., Vandermeulen, J.H., Crowell, M.J. & Patriquin, D.G., 1987. Impact of experimentally dispersed crude oil on vegetation in northwestern saltmarsh - preliminary observations. In: *Proceedings: 1987 Oil Spill Conference (Prevention, Behaviour Control, Cleanup)*. P. Lane & Associates, Halifax, Canada, 509-514.

Larson, D.W., Matthes, U, & Kelly, P.E., 1999. *Cliff Ecology*. Cambridge Studies in Ecology, Cambridge University Press, Cambridge.

Lee, E.M., 1995. Coastal cliff recession in Great Britain: the significance for sustainable coastal management. In: *Directions in Coastal Management*, eds., M.G. Healy & J.P. Doody, Volume 1 of the Proceedings of the 5th EUCC Conference, Swansea, Samara Publishing, Cardigan, 185-194.

Leftic, L., 1995. Coastal areas management programme of the Mediterranean Action Plan of UNEP. In: *Coastal Management and Habitat Conservation*, eds., A.H.P.M. Salman, H. Berends & M. Bonazountas, Proceedings of the 4th EUCC Congress, Vol. **I**, Marathon, Greece, April, 1993. EUCC, Leiden, 3-20.

Liddle, M.J., 1975. A selective review of the ecological effects of human trampling on natural ecosystems. *Biological Conservation*, **7**, 17-36.

Long, S.P. & Mason, C.F., 1983. *Saltmarsh Ecology*. Blackie, Glasgow and London.

Lovric, A.Z., 1993a. Dry coastal ecosystems of Croatia and Yugoslavia In: *Ecosystems of the World 2A. Dry Coastal Ecosystems - Polar Regions and Europe*, ed., E. van der Maarel, Elsevier, The Netherlands, 391-419.

Lovric, A.Z., 1993b. Dry coastal ecosystems of Albania. In: *Ecosystems of the World 2A. Dry Coastal Ecosystems - Polar Regions and Europe*, ed., E. van der Maarel, Elsevier, The Netherlands, 421 - 428.

Lovric, A.Z. & Uslu, T., 1993a. Dry coastal ecosystems of Turkey In: *Ecosystems of the World 2A. Dry Coastal Ecosystems - Polar Regions and Europe*, ed., E. van der Maarel, Elsevier, The Netherlands, 443 - 460.

Lovric, A.Z. & Uslu, T., 1993b. Dry coastal ecosystems of the Black Sea coasts of Bulgaria, Romania and the former Soviet Union. In: *Ecosystems of the World 2A. Dry Coastal Ecosystems - Polar Regions and Europe*, ed., E. van der Maarel, Elsevier, The Netherlands, 475 - 486.

Lundberg, A. (& Losvik, M.C.), 1993. Dry coastal ecosystems of central and southern Norway. In: *Dry Coastal Ecosystems. 2A. Polar Regions and Europe*, ed., E. Van der Maarel, Elsevier, Amsterdam, 119-127.

Lundberg, A., 1996. Changes in the vegetation and management of saltmarsh communities in southern Norway. In: *Studies in European Coastal Management*, eds., P.S. Jones, M.G. Healy & A.T. Williams, Samara Publishing, 197-206.

Macdougall, Z., 1996. The management and protection of internationally famous coastlines: the case of the Venice lagoon. In: *Partnership in Coastal Management*, eds., J. Taussik & J. Mitchell, Samara Publishing, Cardigan, 589-594.

Macey, M.A., 1975. Report 1d. Survey of semi-natural reclaimed marshes. In: *Aspects of the Ecology of the Coastal Area in the Outer Thames Estuary and the Impact of the Proposed Maplin Airport*. Report to DoE by Natural Environment Research Council, Swindon.

Malloch, A.J.C., 1971. Vegetation of the maritime cliff-tops of the Lizard and Land's End peninsulas, west Cornwall. *New Phytologist*, **70**, 155-197.

Marrs, R.H., 1985. Scrub control. In: *Sand Dunes and Their Management*, ed., P. Doody, Focus on nature conservation, **13**, Nature Conservancy Council, Peterborough, 243-251.

Marsden, C., 1947. *The English at the Seaside*, Britain in Pictures, **112**, Collins, London.

Marshal, J.R., 1962. The physiographic development of Caerlaverock Merse. *Transactions of the Dumfries and Galloway Natural History and Antiquarian Society*, **39**, 102-123.

May, V.J., 1977. Earth cliffs. In: *The Coastline*, ed., R.S.K. Barnes, John Wiley & Sons, Chichester, 215-236.

McLachlan, A & Burns, M., 1992. Headland bypass dunes on the South African coast: 100 years of (mis) management. In: *Coastal Dunes: Geomorphology, Ecology and Management: Proceedings of the Third European Dune Congress, Galway, Ireland, 17-21 June 1992*, eds., R.W.G. Carter, T.G.F. Curtis & M.J. Sheehy-Skeffington, A.A. Balkema, Rotterdam, 71-79.

McLusky, D.S., 1971. *Ecology of Estuaries*. Heinemann Educational Books, London.

Meade, C., 1983. *Bird Migration*. Country Life Books, London.

Meiggs, R., 1983. *Trees and Timber in the Mediterranean World*. Oxford University Press.

Mendelssohn, I.A. & McKee, K., 1988. *Spartina alterniflora* die back in Louisiana: time course investigation of soil waterlogging effects. *Journal of Ecology*, **76**, 509-521.

Merino, J., Martin, A., Granados, M. & Merino, O., 1990. Desertification of coastal sands of south-west Spain. *Agriculture, Ecosystems and Environment*, **33**, 171-180.

Merne, O.J., 1991. Birds of Irish dunes - a review. In: *A Guide to the Sand Dunes of Ireland*, ed., M.B. Quigley, European Union for Coastal Conservation, Galway, Ireland, 72-76.

Millard. A.V. & Evans. P.R., 1984. Colonisation of mudflats by *Spartina anglica*: some effects on invertebrates and shore bird populations at Lindisfarne In: *Spartina anglica in Great Britain*, ed., J.P. Doody, Focus on nature conservation, No. **5**, Nature Conservancy Council, Attingham Park, 41-49.

Milliman, J.D., 1996. *Sea-level Rise and Coastal Subsidence: Causes, Consequences, and Strategies*, Kluwer Academic Publishers Group, Boston.

Mitchley, J. & Malloch, J.C., 1991. *Sea Cliff Management Handbook for Great Britain*. University of Lancaster, Lancaster and the Joint Nature Conservation Committee, Peterborough.

Monaghan, P. 1992. Seabirds and sandeels: the conflict between exploitation and conservation in the northern North Sea. *Biodiversity and Conservation*, **1**, 98-111.

Moors P.J. & Atkinson, I.A.E., 1984. Predation on seabirds by introduced animals and factors affecting its severity. In: *Status and Conservation of the World's Seabirds*, eds., J.P. Croxall, P.G.H. Evans & R.W. Schreiber, International Council for Bird Preservation, Technical publication, **2**, 667-690.

Morris, R. & Parsons, M., 1993. Dungeness - a shingle beach and its invertebrates. *British Wildlife*, **4** (3), 137-144.

Musi, F., Perco, F. & Utmar, P., 1992. Loss, restoration and management of wetlands in Friuli - Venezia Giulia, North-eastern Italy. In: *Managing Mediterranean Wetlands and their Birds*, eds., M. Finlayson, T. Hollis & T. Davis. IWRB Special Publication No. **20**, 257-261.

Nakanishi, H., 1982. Coastal vegetation on the shingle spits of south western Japan. *Phytocoenologia*, **10**, 57-71.

Nakanishi, H., 1984. Phytosociological studies on the shingle beach vegetation in central and southern Japan. *Hikobia*, **9**, 137-145.

Natural Environment Research Council, 1998. *LOIS Overview CD-ROM*. Land-Ocean Interaction Study, Centre for Coastal and Marine Sciences, Plymouth.

Nature Conservancy Council, 1989. *Guidelines for the Selection of Biological SSSIs*. Nature Conservancy Council, Peterborough.

Nordberg, L., 1995. Coastal conservation in selected European states. In: *Directions in European Coastal Management*, eds., M.G. Healy & J.P. Doody, Samara Publishing, Cardigan, 47-50.

Nordberg, L., 1999. Coastal legislation and policies in the Baltic sea region. In: *Connecting science and management in the coastal zone*, ed., K. Rabski, Proceedings of the 7th EUCC International Conference, Coastlines '99, European Union for Coastal Conservation, Poland, 28-31.

Nordstrom, K.N., Psuty, N. & Carter, R.W.G. eds., 1990. *Coastal Dunes - Processes and Morphology*. John Wiley & Sons, Chichester.

Norman, F.I., 1971. Predation by the fox (*Vulpes vulpes*) on colonies of the short-tailed shearwater *Puffinus tenuirostris* (temmink) in Victoria Australia. *Journal of Applied Ecology*, **8**, 21-32.

North Sea Task Force, 1993. *North Sea Quality Status Report 1993*. Oslo and Paris Commission, London.

Nowicki, P., Bennett, G., Middleton, D., Rientjes, S. & Walters, R., 1996. *Perspectives on Ecological Networks*. European Centre for Nature Conservation series on Man and Nature, Vol. **1**, Tilburg, the Netherlands.

Nowicki, P., ed., 1998. *The Green Backbone of Central and Eastern Europe*. European Centre for Nature Conservation series on Man and Nature, Vol. **3**, Tilburg, the Netherlands.

Ogilvie, A.G., 1923. The physiography of the Moray Firth coast. *Transactions of the Royal. Society of Edinburgh*, **53**, 377-379.

Oliver, F.W., 1925. *Spartina townsendii*: its modes of establishment. economic uses and taxonomic status, *Journal of Ecology*, **13**, 76-91.

Orford, J.D., Forbes, D.L. & Jennings, S.C., in Press. Origin, development and breakdown of gravel-dominated coastal barriers in Atlantic Canada: future scenarios for North western Europe. In: *Ecology & Geomorphology of Coastal Shingle*, eds., J.R. Packham, R.E. Randall, R.S.K. Barnes & A. Neal. Westbury Academic & Scientific Publishing, Otley, West Yorkshire.

Organisation for Economic Co-operation and Development, 1993a. *Coastal Zone Management. Integrated Policies*, OECD, Paris.

Organisation for Economic Co-operation and Development, 1993b. *Coastal Zone Management. Selected Case Studies*, OECD, Paris.

Oró, D. & Martínez, A., 1994. Migration and dispersal of Audouin's gull *Larus audouinii* from the Ebro Delta colony. *Ostrich*, **65**, 373-392.

Packham, J.R. & Spiers, T., in Press. Plants along the prom: an account of shingle vegetation associated with coastal defence works at Brighton, UK. In: *Ecology & Geomorphology of Coastal Shingle*, eds., J.R. Packham, R.E. Randall, R.S.K. Barnes & A. Neal. Westbury Academic & Scientific Publishing, Otley, West Yorkshire.

Packham, J.R. & Willis, A.J., 1997. *Ecology of Dunes, Saltmarsh and Shingle*. Chapman & Hall, London.

Packham, J.R. & Willis, A.J., in Press. Braunton Burrows in context, a comparative management study. In: *Coastal Dune Management: shared experience of European conservation practice*, eds. J.A. Houston, S.E. Edmondson, and P.J. Rooney, Liverpool University Press.

Palanques, A. & Guillén, J., 1998. Coastal changes in the Ebro Delta: natural and human factors. *Journal of Coastal Conservation*, **4**, 17-26.

Patten, K., ed., 1997. *Second International Spartina Conference*. Washington State University, Long Beach, WA.

Pearce, F., 1993. How green is your golf? *New Scientist*, 30-35.

Percival, S.M., Sutherland, W.J. & Evans, P.R., 1998. Intertidal habitat loss and wildfowl numbers: applications of a spatial depletion model. *Journal of Applied Ecology*, **35**, 57-63.

Pérez, J.J., 1997. Conservation of small islands. *Coastline*, Vol. **6** No. **2**, European Union for Coastal Conservation, Leiden, 12-13.

Perring, F.H., & Farrell, L., 1983. *British Red Data Books: 1: Vascular Plants*. 2nd ed., Royal Society for Nature Conservation, Lincoln.

Perring, F.H. & Walters, S.M., 1976 (2nd Edition). *Atlas of the British Flora*. EP Publishing Ltd., East Ardsley, Wakefield, England.

Perry, B., 1985. *The Middle Atlantic Coast. Cape Hatteras to Cape Cod*. Sierra Club Naturalist's Guide, Sierra Club Books, San Francisco.

Pethick, J., 1984. *An Introduction to Coastal Geomorphology*. Edward Arnold, London.

Pethick, J., 1992. Saltmarsh geomorphology. In: *Saltmarshes - Morphodynamics, Conservation and Engineering Significance*, eds., J.R.L. Allen & K. Pye, Cambridge University Press, 51-62.

Pethick, J.S., 1996. The geomorphology of mudflats. In: *Estuarine Shores: Evolution, Environments and Human Alterations*, eds., K.F. Nordstrom & C.T. Roman, Wiley, Chichester, 185-211.

Petrova, A. & Apostolova, L., 1995. Rare plant species on coastal sand dunes and sand beaches in Bulgaria - status and conservation problems. In: *Coastal Management and Habitat Conservation*, eds., A.H.P.M. Salman, H. Berends & M. Bonazountas. Proceedings of the 4th EUCC Congress, Vol. **I**, Marathon, Greece, April, 1993, European Union for Coastal Conservation, Leiden, The Netherlands, 481-487.

Philp, E.G. & McLean, I.F.G., 1985. The invertebrate fauna of Dungeness. In: *Dungeness: ecology and conservation*, eds., B. Ferry & S. Waters, Focus on nature conservation, **12**, Nature Conservancy Council, Peterborough, 94-135.

Pienkowski, M.W. & Pienkowski, A.E., 1983. WSG project on the movement of wader populations in western Europe, eighth progress report. *Wader Study Group Bulletin*, **38**, 13-22.

Piersma, T. & Davidson, N., eds., 1992. *The Migration of Knots*. Wader Study Group Bulletin, **64**, Supplement, Wader Study Group, Tring.

Pilkey, O.H. & Neal, W.J., series eds., 1984. *Living with the Shore*. Duke University Press, Durham, North Carolina.

Piotrowska, H., 1988. The dynamics of the dune vegetation on the Polish Baltic coast, *Vegetatio*, **77**, 169-175.

Piotrowska, H., 1989. Natural and anthropogenic changes in sand dunes and their vegetation on the southern Baltic coast. In: *Perspectives in Coastal Dune Management*, eds., F. van der Meulen, P.D. Jungerius & J. Visser, SPB Academic Publishing, The Hague, 33-40.

Polunin, O. & Walters, M., 1985. *A guide to the Vegetation of Britain and Europe*. Oxford University Press.

Pritchard, D., 1952. Estuarine hydrology. *Advances in Geophysics*, **1**, 243-280.

Puurmann, E. & Ratas, U., 1995. Problems of conservation and management of the West Estonia seashore meadows. In: *Directions in European Coastal Management*, eds., M. Healy & J.P. Doody, European Union for Coastal Conservation and Samara Publishing, Cardigan, 345-349.

Pye, K., 1996. Evolution of the shoreline of the Dee Estuary, United Kingdom. In: *Estuarine shores, Evolution, Environments and Human Alterations*, eds., K.F. Nordstrom & C.T. Roman, Wiley, 15-34.

Pye, K. & Tsoar, H., 1990. *Aeolian Sand and Sand Dunes*. Unwin Hyman, London.

Quigley, M.B., ed., 1991. *A Guide to the Sand Dunes of Ireland*. European Union for Coastal Conservation, Galway, Ireland.

Radley, G., 1997. Assisting Nature. Landscape Design, *Journal of the Landscape Institute*, **254**, 12-16.

Radley, G.P., 1994. *Sand Dune Vegetation Survey of Great Britain. Part 1 - England.* Peterborough, Joint Nature Conservation Committee.

Randall, R.E., 1977. Shingle formations. In: *The Coastline*, ed., R.S.K. Barnes. John Wiley and Sons, London, 49-62.

Randall, R.E., 1987. *The Status of Mertensia maritima in GB*. Interim Report to the Nature Conservancy Council, Huntingdon.

Randall, R.E., 1992. The vegetation of the coastline of New Zealand: Nelson Boulder bank and Kaitorete Spit. *New Zealand Journal of Geography*, **93**, 11-19.

Randall, R.E., 1996. The shingle survey of Great Britain and its implications for conservation management. In: *Coastal Management and Habitat Conservation*, Vol. **II**, eds., A.H.P.M. Salman, M.J. Langeveld & M. Bonazountas, European Union for Coastal Conservation, Leiden, The Netherlands 360-369.

Randall, R.E. & Doody, J.P., 1995. Habitat Inventories and the European Habitats Directive - the example of Shingle Beaches. In: *Directions in European Coastal Management*, eds., M.G. Healy & J.P. Doody, Samara Publishing, Cardigan, 19-37.

Randall, R.E. & Fuller, R.M., in Press. The Orford shingles, Suffolk, UK: evolving solutions in coastline management. In: *Ecology & Geomorphology of Coastal Shingle*, eds., J.R. Packham, R.E. Randall, R.S.K. Barnes & A. Neal. Westbury Academic & Scientific Publishing, Otley, West Yorkshire.

Ranwell, D.S., 1964a. *Spartina* salt marshes in southern England. II. Rate and seasonal pattern of sediment accretion. *Journal of Ecology*, **52**, 79-95.

Ranwell, D.S., 1964b. *Spartina* saltmarshes in southern England. III, Rates of establishment, succession and nutrient supply at Bridgwater Bay, Somerset. *Journal of Ecology*, **52**, 95-105.

Ranwell, D.S., 1972a. *Ecology of Salt Marshes and Sand Dunes*. Chapman and Hall, London.

Ranwell, D.S., ed., 1972b. *The Management of Sea Buckthorn Hippophae rhamnoides on Selected Sites in Great Britain.* Report of the *Hippophae* Study Group, The Nature Conservancy, Norwich..

Ranwell, D.S., 1967. Worlds resources of *Spartina townsendii* (*sensu lato*) and economic use of *Spartina* marshland. *Journal of Applied Ecology*, **4**, 239-256.

Ranwell, D.S., 1975. The dunes of St. Ouen's Bay, Jersey: an ecological survey. *Annual Bulletin of the Société Jersiaise*, **21**, 381-391.

Ranwell, D.S. & Boar, R., 1986. *Coast Dune Management Guide*. Institute of Terrestrial Ecology, HMSO, London.

Ranwell, D.S. & Downing, B.M., 1960. The use of Dalapon and substituted urea herbicides for control of seed-bearing *Spartina* (cordgrass) in inter-tidal zones of estuarine marsh. *Weeds*, **8**, 78-88.

Ratas, U. & Nilson, E., 1997. *Small Islands of Estonia: Landscape and Ecological Studies*. Ökoloogia Instituut Publication, **5**, Tallin.

Ratas, U. & Puurmann, E., 1995. Human impact on the landscape of small islands in the West-Estonia archipelago. *Journal of Coastal Conservation*, **1** (2), 119-126.

Ratcliffe, D.A., 1977. *A Nature Conservation Review: the Selection of Biological Sites of National Importance to Nature Conservation in Britain.* Vols. 1 & 2, Cambridge University Press, Cambridge.

Reid, A.J, Leach, S.J. & Newlands, C., 1988. *A Botanical Survey of Grazing Marsh Ditch Systems on the North Norfolk Coast.* Unpublished report to the Nature Conservancy Council, England Field Unit, Peterborough.

Regnauld, H. & Kuzucuoglu, C., 1992. Reconstruction of a dune-field landscape after a catastrophic storm: beaches of Ille-et-Vilaine, northern Brittany, France. In: *Coastal Dunes: Geomorphology, Ecology and Management: Proceedings of the Third European Dune Congress, Galway, Ireland, 17-21 June 1992*, eds., R.W.G. Carter, T.G.F. Curtis & M.J. Sheehy-Skeffington, A.A. Balkema, Rotterdam, 379-387.

Repečka R., 1994. Investigation of the fish fauna in the Lithuanian part of the Curonian Lagoon. In: *Coastal conservation and management in the Baltic Region*, eds., V. Guidelis, R. Povilanskas & A. Roepstorff. Proceedings of the EUCC-WWF conference 2-8 May 1994, Riga-Klaipeda-Kaliningrad, 117-123.

Ritchie, A. & Ritchie, G., 1978. *The Ancient Monuments of Orkney*. HMSO, Edinburgh.

Ritchie, W. 1995. Maritime oil spills - environmental lessons and experiences with special reference to low-risk coastlines. *Journal of Coastal Conservation*, **1**, 63-76.

Ritchie, W., 1979. Machair development and chronology in the Uists and adjacent islands. *Proceedings of the Royal Society of Edinburgh*, **77B**, 107-122.

Ritchie, W. & Kingham, L. eds., 1997. *The St. Fergus Coastal Environment, Monitoring and Assessment of the Coastal Dunes at the North Sea Gas Terminals St. Fergus*. Aberdeen University Research & Industrial Services Ltd.

Roberts, B.A. & Robertson, A., 1986. Saltmarshes of Atlantic Canada: their ecology and distribution. *Canadian Journal of Botany*, **65**, 555-567.

Robinson N.A. & Pringle A.W., eds., 1987. *Morecambe Bay, an Assessment of Present Ecological Knowledge*, Morecambe Bay Study Group with the University of Lancaster.

Rodwell, J.S., ed., 2000. *British Plant Communities. Volume 5, Maritime Communities and Vegetation of Open Habitats*. Cambridge University Press, Cambridge.

Rooney, P.J. & Houston, 1998. Management of dunes and dune heaths: experience on the Sefton coast, northwest England. In: *Coastal Dunes, Management, Protection and Research*, ed., C.H. Ovesen, report from a European seminar, Skagen, Denmark, National Forest and Nature Agency, Geological Survey of Denmark and Greenland, 121-129.

Ross, A. Wein, R.S. & Bliss, L.C., 1973. Experimental crude oil spills on arctic plant communities. *Journal of Applied Ecology*, **10**, 671-682.

Rotnicki, K., ed., 1994. *Changes of the Polish Coastal Zone*. IGU Symposium, Quaternary Research Institute, Adam Mackiewiez University, Pozan, Poland.

Royal Commission on Coastal Erosion (and Afforestation). 1911. Third (and final) Report, Vol. **III**, Pt. V. *The reclamation of Tidal lands*.

Rufino, R. & Neves, R., 1992. The effects on wader populations of the conversion of salinas into fish farms. In: *Managing Mediterranean Wetlands and Their Birds*. Proceedings of an IWRB International Symposium, Grado, Italy, eds., M. Finlayson, T. Hollis & T. Davis. IWRB Special Publication, **20**, 177-182.

Sacchi, C.F., 1979. The coastal lagoons of Italy. In: *Ecological Processes in Coastal Environments*, eds., R.L. Jefferies & A.J. Davy, Blackwell Scientific Publications, Oxford, 593-601.

Sadoul, N., Walmsley, J.G. & Charpentier, B., 1998. *Salinas and Nature Conservation*. Conservation of Mediterranean Wetlands, No. **9**, Tour du Valat, Arles (France).

Salisbury, E.J., 1952. *Downs and Dunes*. Bell, London.

Schekkerman, H., Meininger, P.L. & Meire, P.M., 1996. Changes in waterbird populations of the Oosterschelde, SW. Netherlands, as a result of large scale coastal engineering works, *Hydrobiologia*, **282/283**, 509-524.

Schreiber, R.L., Diamond, A.W., Peterson, R.T. & Cronkite, W., 1987. *Save the Birds*. Press Syndicate of the University of Cambridge.

Scott, D.A. & Rose, P.M., 1996. *Atlas of Anatidae populations in Africa and Western Eurasia*. Wetlands International Publication, **41**, Wetlands International.

Scott, G.A.M., 1973. The ecology of shingle beach plants. *Journal of Ecology*, **51**, 517-527.

Scott, G.A.M. & Randall, R.E., 1976. *Crambe maritima* L.: Biological Flora of the British Isles. *Journal of Ecology*, **64**, 1077-1091.

Sell, D , Baker, J.M., Dunnet, D.M., Mcintyre, A.D. & Clark, R.B., 1995. Scientific criteria to optimise oil spill clean up. *Proceedings of the International Oil Spill Conference*, Long Beach California, 595-610.

Shardlow, M., in Press. A review of the conservation importance of shingle habitats for invertebrates in the United Kingdom. In: *Ecology & Geomorphology of Coastal Shingle*, eds., J.R. Packham, R.E. Randall, R.S.K. Barnes & A. Neal. Westbury Academic & Scientific Publishing.

Shennan, I., 1989. Holocene crustal movements and sea level changes in Great Britain. *Journal of Quaternary Science*, **5**, 77-89.

Shuisky, I., 1983. Albania. In: *The World's Coastlines*, eds., E.C.F. Bird & M.L. Schwartz, Stroudsboug, Pennsylvania.

Shuisky, Y., 1995. Erosion and coastal defence on the Black Sea shores. In: *Directions in coastal management*, eds., M.G. Healy & J.P. Doody, Proceedings of the 5th EUCC Conference, Swansea, Samara Publishing, Cardigan, 207-212.

Simmons, I., 1980. Iron Age and Roman coast around the Wash. In: *Archaeology and Coastal Change*, ed., F.H. Thompson. Occasional Paper No. **1**, The Society of Antiquities, London.

Simms, E., 1992. *British Larks, Pipits and Wagtails*. Harper Collins Publishers, London.

Single, M.B. & Hemingsen, M., in Press. Mixed sand and gravel beaches of South Canterbury, New Zealand. In: *Ecology & geomorphology of coastal shingle*, eds., J.R. Packham, R.E. Randall, R.S.K. Barnes & A. Neal. Westbury Academic & Scientific Publishing, Otley, West Yorkshire.

Skarregaard, P., 1989. Stabilisation of coastal dunes in Denmark. In: *Perspectives in Coastal Dune Management*, eds., F. van der Meulen, P.D. Jungerius & J. Visser, SPB Academic Publishing, The Hague, 181-161.

Smith, J.E., ed., 1968. *'Torrey Canyon' Pollution and Marine Life. A Report by the Plymouth Laboratory of the Marine Biological Association of the United Kingdom*. Cambridge University Press, Cambridge.

Smit, C.J. & Wolff, W.J. eds., 1981. *Birds of the Wadden Sea*. Balkema, Rotterdam.

Sneddon P. & Randall, R.E., 1993a. *Coastal Vegetated Shingle Structures of Great Britain: Main Report.* Joint Nature Conservation Committee, Peterborough.

Sneddon, P. & Randall, R.E. 1993b. *Vegetated Shingle Structures Survey of Great Britain: Appendix 1 - Wales*. Joint Nature Conservation Committee, Peterborough.

Sneddon, P. & Randall, R.E. 1994a. *Vegetated Shingle Structures Survey of Great Britain: Appendix 2 - Scotland*. Joint Nature Conservation Committee, Peterborough.

Sneddon, P. & Randall, R.E. 1994b. *Vegetated Shingle Structures Survey of Great Britain: Appendix 3 - England*. Joint Nature Conservation Committee, Peterborough.

Stanley, D.J. & Warne, A.G., 1993. Nile delta: recent geological evolution and human impact. *Science*, **260**, 628-634.

Stapf, 0., 1913. Townsend's grass or ricegrass. *Proceedings Bournemouth Natural Science Society*, **5**, 76-81.

Steers, J.A., 1927. Orford Ness: a study in coastal physiography. *Proceedings of the Geologists' Association*, **37B**, 307-325.

Steers, J.A., ed., 1960. *Scolt Head Island*. Heffer, Cambridge.

Steers, J.A., 1969a. *The Coastline of England and Wales*. 2nd edition. Cambridge University Press, Cambridge.

Steers, J.A., 1969b. *Protection of Coastal Areas*. European Information Centre for Nature Conservation, Strasbourg.

Steers, J.A., 1973. *The Coastline of Scotland*. Cambridge University Press, Cambridge.

Steers, J.A., 1981. *Coastal Features of England and Wales; Eight Essays*. Oleander Press, Cambridge.

Strann, K.-B., 1992. Numbers and distribution of knot *Calidris canutus islandica* during spring migration in north Norway 1983-1989. In: *The migration of knots*, eds., T. Piersma & N. Davidson, Wader Study Group Bulletin, **64**, 121-125.

Sturgess, P., 1992. Clear-felling dune plantations: studies in vegetation recovery In: *Coastal Dunes: Geomorphology, Ecology and Management: Proceedings of the Third European Dune Congress, Galway, Ireland, 17-21 June 1992*, eds., R.W.G. Carter, T.G.F. Curtis & M.J. Sheehy-Skeffington, A.A. Balkema, Rotterdam, 339-349.

Subotowicz, W., 1994. Catastrophical transformation of the cliff coast in Poland. In: *Littoral 94, Proceedings of the Second International Symposium*, eds., Soares de Carvalho & V. Gomes, Vol. **1**, Lisbon, Eurocoast, 281-287.

Sullivan J., 1991. Tiny eels are the best bait for bitterns. *New Scientist*, **112**, 9.

Suter, J.R., 1994. Deltaic coasts. In: *Coastal Evolution - Late Quaternary Shoreline Morphodynamics*, eds., R.W.G. Carter & C.D. Woodroffe, Cambridge University Press, Cambridge, 87-120.

Sutton, P., 1999. The Scaly Cricket in Britain. A complete history from discovery to citizenship. *British Wildlife*, **10**, 145-151.

Tansley, A.G., 1945. *Our Heritage of Wild Nature. A Plea for Organised Nature Conservation.* Cambridge University Press, Cambridge.

Taylor, R.B., Wittman, S.L., Milne, M.J. & Kober, S.M., 1985. *Beach Morphology and Coastal Changes at Selected Sites, Mainland Nova Scotia*. Geological Survey of Canada, Paper, **85-12**.

Teal, J. & Teal, M., 1969. *Life and Death of the Salt marsh*. Ballantine Books, New York.

Tekke, R.M.H., & Salman, A.H.P.M., 1995. Coastal woodlands, forestry and conservation along the Atlantic and North Sea shores. In: *Coastal Management and Habitat Conservation*, eds., A.H.P.M. Salman, H. Berends & M. Bonazountas, Proceedings of the 4th EUCC Congress, Vol. **I**, Marathon, Greece, April, 1993. EUCC, Leiden, 395-409.

Thomas, J.A.,1989. Return of the Large Blue Butterfly. *British Wildlife*, **1** (1), 2-13.

Thompson, H.V. & Worden, A.N., 1956. *The Rabbit*. Collins New Naturalist, London.

Thompson, J.D., 1990. *Spartina anglica*, characteristic feature or invasive weed of coastal salt marshes? *Biologist*, **37**, 9-12.

Thornton, D. & Kite, D.J., 1990. *Changes in the Extent of the Thames Estuary Grazing Marshes*. Internal report, Nature Conservancy Council, London.

Toft, A.R., Pethick, J.S., Burd, F., Gray, A.J., Doody, J.P. & Penning-Rowsell, E., 1995. *A Guide to Understanding and Management of Saltmarshes*. National Rivers Authority, Research & Development Note **324**, Foundation for Water Research, Marlow.

Tooley, M.J., 1993. Long term changes in eustatic sea level. In: *Climate and Sea Level Change - Observations, Projections and Implications*, eds., R.A., Warrick, E.M. Barrow & T.M.L. Wigley, University Press, Cambridge, 81-110.

Tooley, M.J. & Jelgersma, S., 1993. *Impacts of Sea-Level Rise on European Coastal Lowlands*, Blackwell Scientific Publications, Oxford.

Torresani, S., 1989. Historical evolution of the coastal settlement in Italy: Ancient Times. In: *Coastlines of Italy*, ed. P. Fabri, American Society of Civil Engineers, New York, 150-159.

Tubbs, C.R., 1996. Estuary birds - before the counting began. *British Wildlife*, **7**, 226-235.

Tucker, G.M. & Evans, M.I., 1997. *Habitats for Birds in Europe - a Conservation Strategy for the Wider Environment*. BirdLife Conservation Series, No. **6**, BirdLife International, Cambridge, UK.

Turner, R.K., Adger, W.N. & Lorenzoni, I., 1998. *Towards Integrated Modelling and Analysis in Coastal Zones: Principles and Practices*. LOICZ Reports & Studies No. **11**, LOICZ IPO, Texel, The Netherlands.

Turner, R.K., Bateman, I. & Brooke, J.S., 1990. Valuing the benefits of coastal defence: A case study of the Aldeburgh sea defence scheme. In: *Ecological Evaluation and Economic Evaluation*, eds., A. Coker & C. Richards, Flood Hazard Research Centre, Middlesex Polytechnic, 55-81.

UNESCO, 1997. *Methodological Guide to Integrated Coastal Zone Management*. UNESCO, Intergovernmental Oceanographic Commission.

Uslu, T., 1995. Coastal dune afforestation policy in Turkey: environmental aspects. In: *Coastal Management and Habitat Conservation*, eds., A.H.P.M. Salman, H. Berends & M. Bonazountas. Proceedings of the 4th EUCC Congress, Vol. **I**, Marathon, Greece, April, 1993, European Union for Coastal Conservation, Leiden, 113-118.

van der Laan, D., 1985. Changes in the flora and vegetation of the coastal dunes of Voorne (The Netherlands) in relation to environmental changes. In: *Ecology of Coastal Vegetation*, eds., W.G. Beeftink, J. Rozema & A.H.L. Huiskes, *Vegetatio*, **61**, 87-95.

van der Laan, D., van Tongeren, O.F.R., van der Putten, W.H. & Veenbaas, G., 1997. Vegetation development in coastal foredunes in relation to methods of establishing marram grass (*Ammophila arenaria*). *Journal of Coastal Conservation*, **3**, 179-190.

van der Maarel, E., ed., 1993. *Ecosystems of the World 2A. Dry Coastal Ecosystems - Polar Regions and Europe*. Elsevier, The Netherlands.

van der Meulen, F. & van der Maarel, E., 1989. Coastal defence alternatives and nature development perspectives. In *Perspectives in Coastal Dune Management*, eds., F. van der Meulen, P.D. Jungerius & J. Visser, SPB Academic Publishing, The Hague, 183-198.

van der Zande, A., 1995. Nature and recreation: allies or enemies in the rural area? In: *Coastal Dunes Recreation and Planning*, ed., J.M. Drees, European Union for Coastal Conservation, Leiden, 60-63.

van Dijk, H.W.J., 1989. Ecological impact of drinking-water production in Dutch coastal dunes, in *Perspectives in Coastal Dune Management*, eds., F. van der Meulen, P.D. Jungerius & J. Visser, SPB Academic Publishing, The Hague, 163-182.

van Dijk, H.W.J., 1992. Grazing domestic livestock in Dutch coastal dunes: experiments, experiences and perspectives. In: *Coastal Dunes: Geomorphology, Ecology and Management: Proceedings of the third European Dune Congress, Galway, Ireland, 17-21 June 1992*, eds., R.W.G. Carter, T.G.F. Curtis & M.J. Sheehy-Skeffington, A.A. Balkema, Rotterdam, 235-250.

Venner, J., 1977. *The eradication of Hippophae rhamnoides L. from the Braunton Burrows sand dune system*. Senior Warden Project, Nature Conservancy Council, Oxford (Unpublished Thesis).

Vestergaard, P. & Alstrup, V., 1996. Loss of organic matter and nutrients from a coastal dune heath in northwest Denmark caused by fire. *Journal of Coastal Conservation*, **2**, 33-40.

Wallage-Drees, J.M., 1988. *Rabbits in the Coastal Sand Dunes: Weighed and Counted*, Drukkerij Mostert, Leiden.

Walmsley, J.G., 1991. Misappreciated salinas. *Naturopa*, No **67**, Council of Europe, Strasbourg, 28.

Walmsley, J.G., 1994. Un approcio pratico alla gestione ambientale nelle saline del Mediterraneo. In: *La gestione degli ambienti costieri e insulari del Mediterraneo*, MEDMARAVIS, Edizioni del Sole, Alghero, Sardegna, 147-168.

Walmsley, J.G., 1995. A practical approach to wildlife management in Mediterranean salinas. *Coastline*, 1995/1, European Union for Coastal Conservation, Leiden, 21-25.

Walmsley, J.G. & Duncan, P., 1993. Industrial Salinas in the Camargue and the Conservation of Breeding Seabird Populations. In: *Status and Conservation of Seabirds*, Calvia, March 1989, MEDMARAVIS, Grup Balear d'Ornitologia & Sociedad Espagnola de Ornitologia, 285-293.

Walter, H., 1979. *Eleonora's falcon: adaptations to prey and habitat in a social raptor*. The University of Chicago Press, Chicago.

Warrick, R.A., & Oerlemans, J., 1990. Sea level rise. In: *Climate Change: the IPCC Scientific Assessment*, eds., J.T. Houghton, G.J. Jenkins & J.J. Ephraums, Cambridge University Press, Cambridge, 257-281.

Warrick, R.A., 1993. Climate and sea level change: a synthesis. In: *Climate and Sea level Change - Observations, Projections and Implications*, eds., R.A., Warrick, E.M. Barrow & T.M.L. Wigley, University Press, Cambridge, 3-24.

Waters, S., 1985. Vegetation of natural wetlands. In: *Dungeness Ecology and Conservation*, eds., B. Ferry & S. Waters. Focus on Nature Conservation. No. **12**. Nature Conservancy Council, Peterborough, 52-63.

Waters, S.J.P., & Ferry, B.W., 1989. Shingle-based wetlands at Dungeness. In: *Dungeness - the Ecology of a Shingle Beach*, eds., B.W. Ferry, S.J.P. Waters & S.L. Jury. *Botanical Journal of the Linnean Society*, **101**, 59-77.

Way, L., 1987. *A review of Spartina control methods*. Unpublished report, Nature Conservancy Council, Peterborough.

Weber, K. & Hoffman, L., 1970. *Camargue - the Soul of a Wilderness*, George G. Harrap & Co. Ltd, London.

Wecker, M., 1998. *Integrating Biological Control in the Integrated Pest Management Program for Spartina alterniflora in Willapa Bay*. Unpublished report, University of Washington.

Welch, R.C., 1989. Invertebrates of Scottish sand dunes. *Proceedings of the Royal Society of Edinburgh*, **96B**, 267-287.

Westhoff, V., 1985. Nature management in coastal areas of western Europe. *Vegetatio*, **62**, 523-532.

Whatmough, J.A., 1995. Case study. grazing on sand dunes: the re-introduction of the rabbit *Oryctolagus cuniculus* L. to Murlough NNR, Co. Down. In: *The National Trust and Nature Conservation 100 Years On*, eds., D.J. Bullock & H.J. Harvey, Academic Press Ltd., London, 39-43.

White, CP., 1989. *Chesapeake Bay. Nature of the Estuary - a Field Guide*. Tidewater, Maryland.

Whiteside, M.R., 1987. *Spartina* colonisation. In: *Morecambe Bay, an Assessment of Present Ecological Knowledge*, eds., N.A. Robinson & A.W. Pringle, Morecambe Bay Study Group with the University of Lancaster, 118-129.

Wiens, J.A., 1995. Is oil pollution a threat to seabirds? Lessons from the Exxon Valdez oil spill. In: *Threats to Seabirds: Proceedings of the 5th International Seabird Group Conference*, ed., M.L. Tasker, UK. Seabird Group, Sandy, 50-51.

Wilkin, P.J., 1989. *Hirudo medicinalis* at Dungeness. In: *Dungeness - the Ecology of a Shingle Beach*, eds., B.W. Ferry, S.J.P. Waters & S.L. Jury, *Botanical Journal of the Linnean Society*, **101**, 45-57.

Willetts, B.B., 1989. Physics of sand movement in vegetated dune systems. *Proceedings of the Royal Society of Edinburgh*, **96B**, 37-49.

Williams, P.H., 1989. Why are there so many bumble bees at Dungeness? In: *Dungeness - the Ecology of a Shingle Beach*, eds., B.W. Ferry, S.J.P. Waters & S.L. Jury, *Botanical Journal of the Linnean Society*, **101**, 31-44.

Willis, A.J., 1989. Coastal sand dunes as biological systems. *Proceedings of the Royal Society of Edinburgh*, **96B**, 17-36.

Willis, A.J., Folkes, B.F., Hope-Simpson, J.F. & Yemm, E.W., 1959. Braunton Burrows: the dune system and its vegetation. Parts 1 & 2. *Journal of Ecology*, **47**, 1-24 and 249-288.

Woodell, S.R.J., 1989. Cape St Vincent and the Sagres Peninsula, Portugal: important biological sites under threat. *Environmental Conservation*, **16**, 33-39.

World Bank, 1992. *Environmental Review and Environmental Strategy Studies*. Committee of Environmental Protection and Preservation, Tirana, Albania.

Wright, L.D., 1977. Sediment transport and deposition at river mouths: a synthesis. *Geological Society of America*, **88**, 857-868.

Wright, P.J. 1996. Is there a conflict between sandeel fisheries and seabirds? A case study at Shetland. In: *Aquatic Predators and their Prey*, ed., S.P.R. Greenstreet & M.L. Tasker, Fishing News Books, London, 154-165.

Wright, P.J., & Bailey M.C. 1993. *Biology of Sandeels in the Vicinity of Seabird Colonies at Shetland.* Scottish Fisheries Research Report **15**/93, Scottish Office Agriculture, Environment and Fisheries Department, Aberdeen.

Yapp, R.H., Johns, D. & Jones, O.T., 1917. The saltmarshes of the Dovey Estuary. II, The saltmarshes. *Journal of Ecology*, **5**, 65-103.

Zino, F., Heredia, B. & Biscoito, M.J., 1996. Action plan for Zino's petrel (*Pterodroma madeira*). In: B. Heredia, L. Rose & M. Painter, eds., *Globally Threatened Birds in Europe: Action Plans*. Council of Europe and BirdLife International, Strasbourg, France, 33-39.

Index

English names of some commoner plants occurring on or near the coast, mentioned in the text.

Ammophila arenaria, Marram grass
Anacamptis pyramidalis, Pyramidal orchid
Arbutus unedo, Strawberry tree
Armeria maritima, Thrift
Artemisia maritima, Sea wormwood
Arthrocnemum glaucum, Perennial glasswort
Asplenium marinum, Sea spleenwort
Atriplex glabriuscula, Babington's orache
Atriplex pedunculata, Pedunculate sea-purslane
Atriplex portulacoides, Sea-purslane
Beta vulgaris , Sea beet
Brassica oleracea, Sea cabbage
Cakile maritima, Sea rocket
Carex arenaria, Sand sedge
Carex ligerica , Carex sp.
Carpobrotus edulis, Hottentot-fig
Centaurium littorale, Seaside centaury
Cerastium alpinum, Alpine mouse-ear
Cistus palhinhae, Cistus sp.
Cochlearia officinalis ssp. *groenlandica* , Scurvygrass
Corallorhiza trifida, Coral-root orchid
Corynephorus canescens, Grey hair-grass
Crambe maritima , Sea kale
Crithmum maritimum, Rock samphire
Cytisus scoparius, Broom
Dactylorhiza incarnata, Early marsh-orchid
Disphyma crassifolium, Purple dewplant
Dryas octopetala, Mountain avens
Elytrigia atherica, Sea couch
Elytrigia juncea (Elymus farctus), Sand couch
Empetrum nigrum, Crowberry
Ephedra distachya, Joint-pine sp.
Epipactis palustris, Marsh helleborine
Erica arborea, Tree heath
Erica cinerea, Bell heather
Erica tetralix, Cross-leaved heath
Erica vagans, Cornish heath
Erodium cicutarium ssp. *pinnatum*, Stork's-bill
Eryngium maritimum, Sea-holly
Euphorbia paralias, Sea spurge
Festuca ovina, Sheep's fescue
Festuca rubra, Red fescue
Frankenia laevis, Sea-heath
Galium verum var. *maritimum,* Lady's bedstraw
Glaucium flavum, Yellow-horned poppy
Glaux maritima, Sea-milkwort
Honckenya peploides, Sea sandwort
Hippophae rhamnoides, Sea-buckthorn
Jasione crispa ssp. *maritima*, Sheep's-bit
Juncus gerardii, Saltmarsh rush
Juncus maritimus, Sea rush
Juniperus communis, Juniper
Juniperus phoenicea, Phoenician juniper
Lathyrus japonicus, Sea pea
Lavatera arborea, Tree-mallow
Leymus arenarius, Lyme-grass
Ligusticum scoticum, Scots lovage
Limonium bellidifolium, Matted sea-lavender
Limonium binervosum, Rock sea-lavender
Limonium humile, Lax-flowered sea-lavender
Limonium vulgare, Common sea-lavender
Liparis loeselii, Fen orchid
Medicago marina, Sea medick
Mertensia maritima, Oyster-plant
Myrtus communis, Myrtle
Oenanthe lachenalii, Parsley water-dropwort
Oenothera biennis, Common evening-primrose
Ophrys sphegodes, Early spider-orchid
Pancratium maritimum, Sea daffodil
Parapholis incurva, Curved hard-grass
Parapholis strigosa, Hard-grass
Parnassia palustris, Grass-of parnassus
Phoenix canariensis, Canary palm
Phragmites australis, Common reed
Pinus contorta, Lodgepole pine
Pinus nigra var. *maritimum,* Black pine
Pinus pinaster, Maritime pine
Pinus pinea, Stone pine
Pinus sylvestris, Scots pine
Pistacia lentiscus, Mastic tree
Plantago maritima, Sea plantain
Polygonum oxyspermum ssp. *raii*, Ray's knotgrass
Posidonia oceanica, Sea-grass
Primula scotica, Scottish primrose
Prunus spinosa, Blackthorn
Puccinellia maritima, Salt-marsh grass

Quercus coccifera, Kermes oak
Quercus ilex, Holm oak
Rosa rugosa, Japanese rose
Rumex crispus, Curled dock
Ruppia maritima, Beaked tasselweed
Sagina intermedia, Pearlwort
Salicornia europaea agg., Samphire or glasswort
Saxifraga oppositifolia, Purple saxifrage
Scirpus (Bolboschoenus) maritimus, Sea club-rush
Sedum rosea, Roseroot
Silene acaulis, Moss campion
Silene conica ssp. *subconica*, Striated catchfly
Silene nutans, Nottingham catchfly
Silene uniflora, Sea campion
Solidago sempervirens, Seaside goldenrod
Spartina anglica, Common cord-grass
Spartina alterniflora, Smooth cord-grass
Spartina maritima, Small cord-grass
Spartina patens, Soft cord-grass
Spartium junceum, Spanish broom
Spergularia rupicola, Rock sea-spurrey
Suaeda maritima, Annual sea-blite
Suaeda vera, Shrubby sea-blite
Thymus polytrichus, Wild thyme
Tripleurospermum maritimum, Sea mayweed
Ulex gallii, Western gorse
Vicia lathyroides, Spring vetch
Zostera spp., Eel-grass

GPSR Compliance

The European Union's (EU) General Product Safety Regulation (GPSR) is a set of rules that requires consumer products to be safe and our obligations to ensure this.

If you have any concerns about our products, you can contact us on ProductSafety@springernature.com

In case Publisher is established outside the EU, the EU authorized representative is:

Springer Nature Customer Service Center GmbH
Europaplatz 3
69115 Heidelberg, Germany

Batch number: 10761599

Printed by Printforce, the Netherlands